THE FIELD GUIDE TO UN HUMAN ERROR

'This is the kind of message the industry needs to listen to.'
Daniel Maurino, ICAO

'Hurrah! Hurrah! Sidney Dekker has provided us with a delightful book, well written, concluding with a most helpful guide to the understanding of accidents and the appropriate way to investigate them. The lessons will be of great value to everyone.'
Don Norman, author of The Psychology of Everyday Things

'Next time I'm lecturing students, I'll be recommending The Field Guide as required reading! Well done.'
Barry Kirwan, System Safety and Human Error, Eurocontrol, France; co-editor of Changing Regulation: Controlling Risks in Society and Human Factors Impacts in Air Traffic Management

'I have just finished reading The Field Guide. I found it both to be very useful and inspiring at the same time – not necessarily common attributes that occur together!!'
Rob Robson, M.D., Chief Patient Safety Officer, Winnipeg, Canada

'This is probably the most useful of all available books on accident investigation.'
Wayne Perkins, Human Factors Analyst, Maritime New Zealand

'The Field Guide makes a fine contribution to accident investigation methodology.'
International Journal of Applied Aviation Studies

'This is great work – a welcome, practical, refreshing guide to human error using a new lens. A must-read for everyone tired of the "old view" of human error.'
Boyd Falconer, University of New South Wales, Australia

'Neatly summarizes many of the issues during the delivery of CRM training.'
Norman MacLeod, Integrated Team Solutions

'Having read The Field Guide, there is no doubt in my mind about Dekker's ability to push the boundaries of conventional thinking on human factors.'
Bryce Fisher, Manager, Safety Promotion and Education, Transport Canada

'... a very insightful and practical look at the human error side of investigations in any system ... anyone with an interest in aviation safety should make this book part of their collection.'
Journal of Air Transportation

'Congratulations on saying new things to the accident investigation community.'
Neil Johnston, Trinity College Dublin

'... very accessible and satisfying text ... a most welcome and useful primer for those entering the field of investigation and for seasoned professionals seeking refreshment and inspiration.'
The RoSPA Occupational Safety and Health Journal

'The Field Guide *has become a standard text in the field and influential in the patient safety movement. Dekker's work is helping make our clinical laboratory safer for patients.*'
Michael Astion, Editor-In-Chief, Laboratory Errors and Patient Safety

'I think The Field Guide *is the best book I've seen to use not only as a guide for leading investigations, but also for anyone who works in a complex organization to read to help them reframe their understanding of why accidents happen. It is outstanding.*'
Celeste Mayer, Patient Safety Officer, University of North Carolina Health Care System

'I hope the book has a large impact upon accident investigators in all industries. The book focuses upon aviation, but obviously the lesson applies to all.'
Donald A. Norman, Northwestern University, USA and Nielsen Norman Group

'I have every executive in the entire Department of Energy reading The Field Guide *as we speak.*'
Todd Conklin, Los Alamos National Laboratory

'Dekker rocked my world.'
David Devantier, Captain, Boeing 737; CRM instructor

'Congratulations on a well designed book. This has significant potential as an aid for all investigators.'
Hans Willemsen, ATC Quality Assurance Specialist, Airservices, Australia

The Field Guide to Understanding Human Error

SIDNEY DEKKER
Lund University, Sweden

ASHGATE

© Sidney Dekker 2006

All rights reserved. No part of this publication may be reproduced, stored in a retrieval system, or transmitted in any form or by any means, electronic, mechanical, photocopying, recording or otherwise without the prior permission of the publisher.

Sidney Dekker has asserted his right under the Copyright, Designs and Patents Act, 1988, to be identified as the author of this work.

Published by
Ashgate Publishing Limited
Gower House
Croft Road
Aldershot
Hampshire GU11 3HR
England

Ashgate Publishing Company
Suite 420
101 Cherry Street
Burlington, VT 05401-4405
USA

Ashgate website: http://www.ashgate.com

British Library Cataloguing in Publication Data
Dekker, Sidney
   The field guide to understanding human error
   1.Industrial safety 2.Human engineering 3.System failures (Engineering)
   I.Title
   363.1'1

Library of Congress Cataloging-in-Publication Data
Dekker, Sidney.
   The field guide to understanding human error / by Sidney Dekker.
      p. cm
   Rev. ed. of: The field guide to human error investigations / Sidney Dekker. 2002.
   Includes bibliographical references and index.
   ISBN 0-7546-4825-7 (hardback) -- ISBN 0-7546-4826-5 (pbk)
1. System failures (Enginering) 2. Human engineering. 3. Industrial accidents. I. Dekker, Sidney. Field guide to human error investigations. II. Title.

   TA169.5.D45 2006
   620.8--dc22

2006005995

ISBN: 978-0-7546-4825-3 (HBK)
ISBN: 978-0-7546-4826-0 (PBK)

Reprinted 2006, 2008

Printed in Great Britain by TJ International Ltd., Padstow, Cornwall.

Contents

List of Figures and Tables		vi
Acknowledgments		ix
Preface		x
1	The Bad Apple Theory	1
2	The New View of Human Error	15
3	The Hindsight Bias	21
4	Put Data in Context	29
5	"They Should Have …"	39
6	Trade Indignation for Explanation	45
7	Sharp or Blunt End?	59
8	You Can't Count Errors	65
9	Cause is Something You Construct	73
10	What is Your Accident Model?	81
11	Human Factors Data	93
12	Build a Timeline	101
13	Leave a Trace	119
14	So What Went Wrong?	135
15	Look into the Organization	159
16	Making Recommendations	173
17	Abandon the Fallacy of a Quick Fix	183
18	What about People's Own Responsibility?	195
19	Making Your Safety Department Work	205
20	How to Adopt the New View	215
21	Reminders for in the Rubble	225
Index		231

List of Figures and Tables

Figures

3.1 Hindsight changes how we look at past decision making. It turns real, convoluted complexity into a simple, linear story; a binary decision to err or not to err. Idea for image by Richard Cook — 25

3.2 Different perspectives on a sequence of events: Looking from the outside and hindsight you have knowledge of the outcome and dangers involved. From the inside, you may have neither — 26

4.1 Micromatching can mean that you take performance fragments from the stream of events and hold them up against rules or procedures that you deem applicable in hindsight. You don't explain anything by doing this — 31

4.2 Cherry-picking means taking fragments from all over the record and constructing a story with them that exists only in your hindsight. In reality, those pieces may have had nothing to do with each other — 35

4.3 It is easy to gather cues and indications from a sequence of events and lob them together as in a shopping bag. This is not, however, how people inside the unfolding situation saw those cues presented to them — 37

4.4 See the unfolding world from the point of view of people inside the situation—not from the outside or from hindsight — 38

5.1 Counterfactuals: Going back through a sequence, you wonder why people missed opportunities to direct events away from the eventual outcome. This, however, does not explain failure — 41

5.2 Judgmental: by claiming that people should have done something they didn't, or failed to do something they should have, you do not explain their actual behavior — 43

6.1 Different perspectives on a sequence of events: Looking from the outside and hindsight you have knowledge of the outcome and dangers involved. From the inside, you may have neither — 47

List of Figures and Tables vii

6.2 The language you use in your description of human error gives away where you stand. Here you clearly take the position of retrospective outsider   52
6.3 Using the right language, you can try to take the perspective of the people whose assessments and actions you are trying to understand   55
7.1 Failures can only be understood by looking at the whole system in which they took place. But in our reactions to failure, we often focus on the sharp end, where people were closest to causing or potentially preventing the mishap   60
8.1 The mirage on the horizon. We think all we need to do to reach perfect safety is get rid of the last 70 per cent human errors. But opportunities for safety are right here, right under our feet. Original metaphor by Richard Cook   66
10.1 Laying out a sequence of events, including people's assessments and actions and changes in the process itself   85
10.2 We may believe that blocking a known pathway to failure somewhere along the way will prevent all similar mishaps   85
10.3 Without understanding and addressing the deeper and more subtle vulnerabilities that surround failure, we leave opportunities for recurrence open   87
10.4 The "Swiss Cheese" analogy. Latent and active failures are represented as holes in the layers of defense. These need to line up for an accident to happen (after Reason, 1990)   89
12.1 Remember at all times to try to see the unfolding world from the point of view of people inside the situation—not from the outside or from hindsight   102
12.2 Connecting critical process parameters to the sequence of people's assessments and actions and other junctures.   112
12.3 Laying out the various (overlapping) tasks that people were accomplishing during an unfolding situation   114
13.1 The interesting cognitive and coordinative dynamics take place *beneath* the large psychological label. The label itself explains nothing   122
13.2 Don't make a leap of faith, from your facts to a big label that you think explains those facts. Leave an analytic trace that shows how you got to your conclusion   124
13.3 Leaving a trace bottom up. Overlapping talk, no response when one is expected, unequal turns at talking and offering repair of

somebody else's talk when none is needed together could point to a "loss of effective CRM" — 128

13.4 The way to bridge the gap between facts and conclusions (about those facts) from the top down is to find a definition in the literature for the phenomenon you suspect is at play, and start looking in your facts for evidence of it — 130

13.5 Leaving a trace top-down. Using a definition for "loss of effective CRM" that lists misunderstanding the problem, no common goal and uncoordinated corrective actions, you can find evidence for that in your facts — 131

14.1 We make assessments about the world, which update our current understanding. This directs our actions in the world, which change what the world looks like, which in turn updates our understanding, and so forth (figure is modeled on Neisser's perceptual cycle) — 136

15.1 At a particular moment in time, behavior that does not live up to some standard may look like complacency or negligence. But deviance may have become the new norm across an entire operation or organization — 162

15.2 Murphy's law is wrong. What can go wrong usually goes right, and over time organizations can come to think that a safety threat does not exist or is not so bad — 165

16.1 The trade-off between recommendations that will be easier to implement and recommendations that will actually have some lasting effect — 176

21.1 If you want to understand human error, see the unfolding world from the point of view of people inside the situation—not from the outside or from hindsight — 225

Tables

0.1 Two views on human error — xi
9.1 Two statements of cause about the same accident — 76
14.1 Finding a mismatch between problem demands and coping resources can help you make arguments about stress and workload more specific. Remember that people feel stress when they perceive a mismatch — 141
14.2 Opposing models of procedures and safety — 157
15.1 Two images of disaster relief work — 167

Acknowledgments

I keep writing books about human error, even though I don't believe that "human error" actually exists—other than as a convenient but misleading explanatory construct; as an intervention in history that helps us structure and make sense of the past. Initially, such a construct may make our own life easier. But it quickly troubles our ability to really understand sources of safety and risk in our own organizations and elsewhere. This is not just my insight, as I am but a student of a set of people and ideas far greater than myself. Without them, neither this Field Guide nor any of my other writings would even exist or be worth reading. The ideas of David Woods, Erik Hollnagel, Nancy Leveson, John Flach, Richard Cook and Jens Rasmussen are among the most prominent ones in this book.

The Field Guide to Understanding Human Error is a successor to, and extension of, *The Field Guide to Human Error Investigations*. Neither this Field Guide, nor its predecessor, nor my other books, would have been written if it hadn't been for the kind of sponsoring from Arne Axelsson, the immediate past Director of Flight Safety in Sweden. With his support throughout the years, he has always understood that scientists need the funding and freedom to let their imagination roam. Such support, on which scientific innovation depends, appears increasingly scarce today.

I am grateful to my editor Guy Loft for buying into this project with faith and enthusiasm and the friends, colleagues and students who have taken the trouble to review half-baked versions of various chapters. I also want to thank readers of the preceding Field Guide for encouraging me to try to give them more, and for their many suggestions for improvements. The result, as imperfect and tentative as any set of ideas out there, is in your hands now.

Sidney Dekker
January 2006

Preface

So you are faced with a human error problem.
What do you do?
How do you make sense of other people's puzzling assessments and actions? How can you get people in your organization to stop making errors? How can you get your operations to become safer?
You basically have two options, and your choice determines the focus, questions, answers and ultimately the success of your efforts, as well as the potential for progress on safety in your organization:

- You can see human error as **the cause of a mishap**. In this case "human error", under whatever label—loss of situation awareness, procedural violation, regulatory shortcomings, managerial deficiencies—is the conclusion of your efforts to understand error.
- You can see human error as **the symptom of deeper trouble**. In this case, human error is the starting point for your efforts. Finding "errors" is only the beginning. You will probe how human error is systematically connected to features of people's tools, tasks and operational/organizational environment.

The first is called the Old View of human error, while the second—itself already 50 years in the making—is the New View of human error. This Field Guide is the successor to the *Field Guide to Human Error Investigations*. It helps you understand human error according to the New View. Whether you are an investigator, a manager, a regulator, a practitioner, the New View can give you new and innovative leverage over your "human error problem". Leverage you may not have known existed.

Embracing the New View is not easy. It will take work. And maybe a change in your own worldview. But embracing the New View is necessary if you really want to create progress on safety.

We have long searched for ways to limit human variability in—what we think are—otherwise safe systems. Performance monitoring, error counting and

Table 0.1 Two views on human error

The Old View of human error on what goes wrong	The New View of human error on what goes wrong
Human error is a cause of trouble	Human error is a symptom of trouble deeper inside a system
To explain failure, you must seek failures (errors, violations, incompetence, mistakes)	To explain failure, do not try to find where people went wrong
You must find people's inaccurate assessments, wrong decisions, bad judgments	Instead, find how people's assessments and actions made sense at the time, given the circumstances that surrounded them

The Old View of human error on how to make it right	The New View of human error on how to make it right
Complex systems are basically safe	Complex systems are not basically safe
Unreliable, erratic humans undermine defenses, rules and regulations	Complex systems are trade-offs between multiple irreconcilable goals (e.g. safety and efficiency)
To make systems safer, restrict the human contribution by tighter procedures, automation, supervision	People have to create safety through practice at all levels of an organization

categorizing—these activities all assume that we can maintain our safety by keeping human performance within prespecified boundaries. Our investigations into human error often reveal how people create havoc in otherwise safe systems when they go outside those boundaries. When people don't do what they are supposed to do. When they violate rules or lose situation awareness.

In fact, while we can make our systems safer and safer, the human contribution to trouble remains stubbornly high (70 per cent!). We have long put our hopes for improving safety on tightening the bandwidth of human performance even further. We introduce more automation to try to get rid of unreliable people. We write additional procedures. We reprimand errant operators and tell them that their performance is "unacceptable". We train them some more. We supervise them better, we tighten regulations.

Those hopes and ideas are now bankrupt. People do not come to work to do a bad job. Safety in complex systems is not a result of getting rid of people, of reducing their degrees of freedom. Safety in complex systems is *created by people through practice*—at all levels of an organization. It's only people who can hold together the patchwork of technologies and tools and do real work in environments where multiple irreconcilable goals compete for their attention (efficiency, safety, throughput, comfort, financial bottom line).

The New View embodies this realization and lays out a new strategy for understanding safety and risk on its basis. Only by understanding the New View can you and your organization really begin to make progress on safety. And the Field Guide is here to help you do just that.

Here is how.

Chapter 1. The Bad Apple Theory
Presents the Old View of human error: unreliable people undermine basically safe systems. In investigations, we must find people's shortcomings and failings. And in efforts to improve safety, we must make sure people do not contribute to trouble again (so, more rules, more automation, more reprimands).

Chapter 2. The New View of Human Error
Explains how human error is a symptom of trouble (engineered, organized, social, etc.) deeper inside the system, and that efforts to understand error begin with seeing how people try to create safety through their practice of reconciling multiple goals in complex, dynamic settings.

Chapter 3. The Hindsight Bias
Presents research on the hindsight bias, one of the best documented biases in psychology and an unwitting foundation of the Old View. Shows how pervasive the effects of hindsight are, and how they interfere profoundly with your ability to understand human behavior that preceded a bad outcome.

Chapter 4. Put Data in Context
Tells you how to avoid the hindsight bias by not mixing your reality with the one that surrounded other people. You have to disconnect your understanding

of the true nature of the situation (including its outcome) from the unfolding, incomplete understanding of people at the time.

Chapter 5. "They Should Have …"
Lays out what counterfactual reasoning is and how it muddles your ability to understand why people did what they did. Sensitizes you to the language of counterfactuals and how it easily slips into investigations of, and countermeasures against, human error.

Chapter 6. Trade Indignation for Explanation
Explains how you can avoid the traps of counterfactual reasoning and judgmental language, and how to move instead to explanations of why behavior made sense to people at the time.

Chapter 7. Sharp or Blunt End?
Shows you how easy it is to revert to proximal explanations of failure by relying on the (in)actions of those closest in time and place to the mishap or to potentially preventing it.

Chapter 8. You Can't Count Errors
Explains how getting a grip on your human error problem does not mean quantifying it. Error categorization tools look unsuccessfully for simple answers to the sources of trouble and sustain the myth of a stubborn 70% human error. This also makes artificial distinctions between human error and mechanical failure.

Chapter 9. Cause is Something You Construct
Talks about the difficulty of pinpointing *the* cause (proximal or root or probable cause) of an accident. Asking what is *the* cause, is just as bizarre as asking what is *the* cause of not having an accident. Accidents have their basis in the real complexity of the system, not their apparent simplicity.

Chapter 10. What is Your Accident Model?
What can count as "cause" depends on the accident model you apply (e.g. sequential, epidemiological, systemic). Some are better for some purposes than others, both when it comes to understanding error and making progress on safety.

Chapter 11. Human Factors Data
Describes some sources of, and some processes for getting at, data relevant to understanding human error and other human factors issues.

Chapter 12. Build a Timeline
Shows how the starting point of understanding error is often the construction of a detailed timeline. Talks about the traps inherent in building a timeline for human performance and how to correct them.

Chapter 13. Leave a Trace
Talks about why labeling human error (under whatever guise) as cause is easily done. Discusses "folk models" that often enter into explanations. Encourages you to not make "leaps of faith" and leave an analytic trace for your conclusions.

Chapter 14. So What Went Wrong?
Offers alternatives to "human error", explaining human performance issues such as breakdowns in coordination, cognitive lock-up, automation surprises, plan continuation, distortion of time perception under stress, and buggy or inert knowledge.

Chapter 15. Look into the Organization
Presents ways to analyze organizational issues behind the creation of human error, such as production pressures, drifting into failure, different images of work, politics and safety culture.

Chapter 16. Writing Recommendations
Offers directions for the writing of good, useful human factors recommendations that convert diagnosis of what went wrong into change (i.e. continuous improvement of your organization).

Chapter 17. Abandon the Fallacy of a Quick Fix
Discusses how failures and investigations are opportunities for learning—if indeed seen that way. Also covers obstacles to learning from failure and tells you not to get deluded by the fallacy of a quick fix. "Human error" problems are organizational problems, and so at least as complex as your organization.

Chapter 18. What About People's own Responsibility?
Presents how you have a choice to seek explanations for failure in individual actors or in the system that helps determine their performance. Covers some

of the factors that fuel debate around this choice. Introduces "no responsibility without proof of authority".

Chapter 19. Making Your Safety Department Work
Talks about the prerequisites of a meaningful safety department (involved, independent, informative and informed) and how safety work is not just bottom-up provision of information but also guidance of top-down countermeasures.

Chapter 20. How to Adopt the New View
Offers much-needed guidance on how you can help your own organization adopt the New View, and guage where your organization is in its growth towards better learning from failure.

Chapter 21. Reminders for in the Rubble
Wraps together the most important lessons of the Field Guide in a number of ideas and steps for you to follow when trying to understand human error.

1 The Bad Apple Theory

There are basically two ways of looking at human error. The first view is known as the Old View, or The Bad Apple Theory. It maintains that:

- Complex systems would be fine, were it not for the erratic behavior of some unreliable people (Bad Apples) in it;
- Human errors cause accidents: humans are the dominant contributor to more than two thirds of them;
- Failures come as unpleasant surprises. They are unexpected and do not belong in the system. Failures are introduced to the system only through the inherent unreliability of people.

This chapter is about the first view. The second is about a contrasting view, known as the New View. The rest of the book helps you avoid the Old View and apply the New View to your understanding of human error.

Every now and again, nationwide debates about the death penalty rage in the United States. Studies find a system fraught with vulnerabilities and error. Some states halt proceedings altogether; others scramble to invest more in countermeasures against executions of the innocent.

The debate is a window on people's beliefs about the sources of error. Says one protagonist: "The system of protecting the rights of accused is good. It's the people who are administering it who need improvement: The judges who make mistakes and don't permit evidence to be introduced. We also need improvement of the defense attorneys."[1] *The system is basically safe, but it contains bad apples. Countermeasures against miscarriages of justice should focus on them. Get rid of them, retrain them, discipline them.*

But what is the practice of employing the least experienced, least skilled, least paid public defenders in many death penalty cases other than systemic? What are the rules for judges' permission of evidence other than systemic? What is the ambiguous nature of evidence other than inherent in a system that often relies on eyewitness accounts to make or break a case?

The Old View maintains that safety problems are the result of a few Bad Apples in an otherwise safe system. These Bad Apples don't always follow the rules, they don't always watch out carefully enough. They undermine the organized and engineered system that other people have put in place so carefully. This, for instance, is what some think creates safety problems in the nuclear power industry:

> Although many of the human frailties and other deficiencies that lie behind the majority of remaining accidents "have been anticipated in safety rules, prescriptive procedures and management treatises, *people don't always do what they are supposed to do*. Some employees have negative attitudes to safety which adversely affect their behaviors. This undermines the system of multiple defences that an organization constructs ..."[2]

This quote (with the emphasis on "some people" actually in the original) embodies all of the tenets of the Old View. Here they are:

- Human frailties lie behind the majority of remaining accidents. Human errors are the dominant cause of remaining trouble that hasn't been engineered or organized away yet.
- Safety rules, prescriptive procedures and management treatises are supposed to control this element of erratic human behavior.
- But this control is undercut by unreliable, unpredictable people who still *don't do what they are supposed to do*.
- Some Bad Apples keep having negative attitudes toward safety, which adversely affects their behavior. So not attending to safety is a personal problem, a motivational one, an issue of mere individual choice.
- The basically safe system, of multiple defenses carefully constructed by the organization, is undermined by erratic people. All we need to do is protect it better from their vicissitudes.

This view, the Old View, and all it stands for, is deeply counterproductive. It has been tried for ages, without noticeable effect. Real progress on safety comes instead from abandoning the idea that errors are causes, and that people are the major remaining threat to otherwise safe systems. Real progress comes from embracing the New View.

Bad People in Safe Systems, or Well-intentioned People in Imperfect Systems?

At first sight, stories of error seem so simple:

- Somebody did not pay enough attention;
- If only somebody had recognized the significance of this indication, of that piece of data, then nothing would have happened;
- Somebody should have put in a little more effort;
- Somebody thought that making a shortcut on a safety rule was not such a big deal.

Given what you know after-the-fact, most errors seem so preventable. In some sense, errors seem the result of a Bad Apple. You wonder how you can cope with the unreliability of the human element (for example, deficient judges) in your system. But such apparent simplicity is misleading. Underneath every seemingly obvious, simple story of error, there is a second, deeper story. A more complicated story. This second story is inevitably an organizational story, a story about the system in which people work:

Underneath every simple, obvious story about error, there is a deeper, more complex story

- Safety is never the only goal. Organizations exist to provide goods or services and to make money at it;
- People do their best to reconcile different goals simultaneously (e.g. service or efficiency versus safety);
- A system isn't automatically safe: people actually have to create safety through practice at all levels of the organization;
- Production pressure influences people's trade-offs, making normal or acceptable what previously was irregular or unsafe;
- New tools or technology that people have to work with, change error opportunities and pathways to failure.

The second story, in other words, is a story of the real complexity in which people work. Not a story about the apparent simplicity.
Second stories of error reveal how people actually have to create success and safety. Systems are not basically safe themselves. These systems are themselves inherent contradictions between operational efficiency on the one hand and safety (for example: protecting the rights of the accused) on the other.

People in these systems learn about the pressures and contradictions, the vulnerabilities and pathways to failure. And they develop strategies to not have failures happen. But these strategies may not be completely adapted. They may be outdated. People may be focusing on the wrong things, the wrong risks.

They may be thwarted by their rules, or by the feedback they get from their management about what "really" is important (often: production, efficiency).

This is called the "New View" of human error. In this view, errors are symptoms of trouble deeper inside a system. Errors are the other side of people trying to pursue success in an uncertain, resource-constrained world. The old view, or the Bad Apple Theory, sees systems as basically safe and people as the major source of trouble. The new view, in contrast, understands that systems are not basically safe. It understands that safety needs to be created through practice, by people.

Take Your Pick: Blame Human Error *or* Try to Learn from Failure

So you can see human error as a cause of trouble in otherwise safe systems. In this case you stop looking any further as soon as you have found a convenient "human error" to blame for the trouble. Such a conclusion and its implications supposedly get to the causes of system failure. Or you can see human error as a symptom of trouble in a system that is not basically safe. Then you will begin to understand that human error is a structural by-product of people trying to pursue success in resource-constrained, uncertain, imperfect systems.

Is human error a cause or a symptom of trouble? The choice is yours

The old and the new view are two drastically different views on what makes systems safe, and how they occasionally fail. If you see error as a symptom, you commit to a radically new perspective on what makes systems safe or risky. Error is a starting point, not a conclusion. You will join an ever larger group of people, ranging from researchers, practitioners, managers and even regulators, who have discovered that efforts to make progress on safety begin with calling off the hunt for human error. "Human error" is not an explanation for trouble. It demands an explanation.

This Field Guide will help you make a commitment to the New View. It will equip you better for applying the New View as you and your organization are looking for ways to recover from a mishap, to stay safe, or become even safer.

Investigations and the Old View

Like debates about human error, investigations into human error mishaps face a choice. A choice between the Bad Apple Theory in one of its many versions, or the New View of human error.

A Boeing 747 Jumbo Jet crashed when taking off from a runway that was under construction and being converted into a taxiway. The weather at the time was bad—a typhoon was about to hit the country: winds were high and visibility low. The runway under construction was close and parallel to the intended runway, and bore all the markings, lights and indications of a real runway. This while it had been used as a taxiway for quite a while and was going to be officially converted at midnight the next day—ironically only hours after the accident.

Pilots had complained about potential confusion for years, saying that it was "setting a trap for a dark and stormy night". Moreover, at the departure end there was no sign that the runway was under construction. The first barrier stood a kilometer down the runway, and behind it a mass of construction equipment—all of it hidden in mist and heavy rain. The chief of the country's aviation administration, however, claimed that "runways, signs and lights were up to international requirements" and that "it was clear that human error had led to the disaster". So human error was simply the cause. There was no deeper trouble of which the error was a symptom.

The ultimate goal of an investigation is to learn from failure. The road that most investigations follow is paved with intentions to pursue the New View. Investigators intend to find the systemic vulnerabilities behind individual errors. They want to address the error-producing conditions that, if left in place, will repeat that pattern of failure.

In practice, however, investigations often return disguised versions of the Bad Apple Theory, both in findings and recommendations. They sort through the rubble of a mishap to:

- Single out particularly ill-performing practitioners;
- Find evidence of erratic, wrong or inappropriate behavior;
- Bring to light people's bad decisions; their inaccurate assessments; their deviations from written guidance or procedures.

Investigations often end up concluding how front-line operators failed to notice certain data, or did not adhere to procedures that appeared relevant only after the fact. If this is what they conclude, then it is logical to recommend the retraining of particular individuals; the tightening of procedures or oversight.

Of course, you might as well not spend the resources on an investigation in this case. After all, you can easily write the conclusions and recommendations without any substantive knowledge of the event. You can even write them *before* the event! All you have to say is: "human error" (by whatever fancy label you deem legitimate: lack of awareness, poor judgment). And then you call for retraining, you issue a reprimand, you send out a reminder that people should follow applicable procedures.

If you conclude "human error", you may as well not have spent money on the investigation

Investigative bodies and departments across the world seem to keep a stack of these stand-bys on the shelf, ready for use whenever they run out of imagination, investigative acumen or political capital to probe any deeper. If you find that you only write such things in your recommendations, you have wasted everybody's time and money. There are, of course, reasons why investigations regress into the Bad Apple Theory.

For example:

- Resource constraints on investigations. Findings may need to be produced in a few months' time, and money is limited;
- Reactions to failure, which make it difficult not to be judgmental about seemingly bad performance;
- The hindsight bias, which confuses our reality with the one that surrounded the people we investigate;
- Political distaste of deeper probing into sources of failure, which may de facto limit access to certain data or discourage certain kinds of recommendations;
- Limited human factors knowledge on part of investigators. While wanting to probe the deeper sources behind human errors, investigators may not really know where or how to look.

In one way or another, The Field Guide will try to deal with these reasons. It also presents approaches for how to do a New View human error investigation.

Making Progress on Safety and the Old View

In the old view of human error, progress on safety is driven by one unifying idea: unreliable people undermine otherwise safe systems.

Charges were brought against the pilots who flew a VIP jet with a malfunction in its pitch control system (which makes the plane go up or down). Severe oscillations during descent killed seven of their unstrapped passengers in the back. Significant in the sequence of events was that the pilots "ignored" the relevant alert light in the cockpit as a false alarm, and that they had not switched on the "fasten seatbelt" sign from the top of descent, as recommended by the procedures. The pilot oversights were captured on video, shot by one of the passengers who died not much later. The pilots, wearing seatbelts, survived the upset.[3]

When you believe that systems are basically safe, you want to protect them from the vagaries of human behavior. Progress on safety, then, supposedly comes from:

- Making sure that defective practitioners (the Bad Apples) do not contribute to system breakdown again. Put them on "administrative leave"; demote them to a lower status; educate or pressure them to behave better next time; instill some fear in them and their peers by taking them to court or reprimanding them.
- Tightening procedures and close regulatory gaps. This reduces the bandwidth in which people operate. It leaves less room for error.
- Introducing more technology to monitor or replace human work. If machines do the work, then humans can no longer make errors doing it. And if machines monitor human work, they can snuff out any erratic human behavior.

But let's see where this gets you.

Adding more procedures

Adding or enforcing existing procedures does not guarantee compliance. A typical reaction to failure is procedural overspecification—patching observed holes in an operation with increasingly detailed or tightly targeted rules that respond specifically to just the latest incident. But procedural overspecification is likely to widen the gap between procedures and practice, rather than narrow it. Rules will grow more and more at odds with the context-dependent nature of practice.

Mismatches between written guidance and operational practice always exist. Think about the work-to-rule strike, a form of industrial action. Workers say: "Let's follow all the rules for a change!" Systems come to a grinding halt. Gridlock is the result. Follow the letter of the law, and the work will not get done. It is as good as, or better than, going on strike.

> **There is always a mismatch between rules and practice. Do you want to increase that by writing more?**

Seatbelt sign on from top of descent in a VIP jet? The layout of furniture in these machines and the way in which their passengers are expected to make good use of their time by meeting, planning, working, discussing, could well discourage people from strapping in any earlier than strictly necessary. Pilots can blink the light all they want, over time it may become pointless to switch it on from 41 000 feet on down.

And who typically employs the pilot of a VIP jet? The person in the back. So guess who can tell whom what to do. And why have the light on only from the top of descent? This is hypocritical—only in the VIP jet upset discussed here was that relevant because loss of control occurred during descent. But other incidents with turbulence and in-flight deaths have occurred during cruise. Procedures are insensitive to this kind of natural variability.

New procedures can also get buried in masses of regulatory paperwork. Mismatches between procedures and practice grow not necessarily because of people's conscious non-adherence, but because of the amount and increasingly tight constraints of procedures.

The vice president of a large airline commented recently how he had seen various of his senior colleagues retire over the past few years. Almost all had told him how they had gotten tired of updating their aircraft operating manuals with new procedures that came out—one after the other—often for no other reason than to close just the next gap that had been revealed in the latest little incident. Faced with a growing pile of paper in their mailboxes, they had just not bothered. Yet these captains all retired alive and probably flew very safely during their last few years.

Procedures are a problematic issue. Their role is often misunderstood and that of "violations" almost always overestimated. More will be said in Chapters 14 and 15.

Adding a bit more technology

We often think that adding just a little bit more technology will help remove human error. After all, if there is technology to do the work, or to monitor the human doing the work, then we have nicely controlled the potential for error. But more technology does not remove the potential for human error. It merely relocates or changes it.

A warning light does not solve a human error problem, it creates new ones. What is this light for? How do we respond to it? What do we do to make it go away? It lit up yesterday and meant nothing. Why respond to it today?

A warning light is just a threshold crossing device: it starts blinking when some electronic or electromechanical threshold is exceeded. If particular values stay below the threshold, the light is out. If they go above, the light comes on. But what is its significance? After all, the aircraft has been flying well and behaving normally, even with the light on. Of course, a warning light is "new technology" only in the most rudimentary sense. Promises and problems of more advanced technology and automation are discussed in Chapter 14.

New technology does not remove human error. It changes it

Removing Bad Apples

Throwing out the Bad Apples, lashing out at them, telling them you are not happy with their performance, may seem like a quick, nice, rewarding fix. But it is like peeing in your pants. It gets nice and warm for a little while, and you feel relieved. But then it gets cold and uncomfortable, and you look like a fool. Lashing out at supposed Bad Apples, at the putative culprits behind all the trouble, is actually a sign of weakness. It shows that you could be at a loss as to what to do in the wake of failure. You actually have no idea how to really make progress on safety. And by bearing down on supposed Bad Apples, you can actually make things a lot worse:

Reprimanding "Bad Apples" is like peeing in your pants. You feel warm and relieved first, but soon you look like a fool

- You fool yourself and your stakeholders (customers, regulator, other employees, the media) that you have done something about the problem;

- You actually haven't done anything to remove the problem that exhibited itself through those people. It leaves the trap in place for the next practitioners. And it leaves you as exposed as you were the first time;
- With fear of punishment, people will hide evidence of mistakes. They will no longer report irregularities. They will remain silent about problems, incidents, occurrences.

So when you think you are "setting an example" with a robust response to a supposed "Bad Apple", think about what you are setting an example for. You will condition your people to shut up, to conceal difficulties. If you hunt down individual people for system problems, you will quickly drive real practice underground. You will find it even more difficult to know how work really takes place. Do you want to wait for the accident to reveal the true picture? Organizations do this all too often, at their own peril.

As it turns out, the VIP jet aircraft had been flying for a long time with a malfunctioning pitch feel system ("Oh that light? Yeah, that's been on for four months now"). These pilots inherited a systemic problem from the airline that operated the VIP jet, and from the organization charged with its maintenance.

Why is the Bad Apple Theory Popular?

Cheap and easy

So why would anyone adhere to the Bad Apple Theory of human error? There are many reasons. It is a deviously straightforward approach to dealing with safety. It is simple to understand and simple, and relatively cheap, to implement. The Bad Apple Theory suggests that failure is an aberration, a temporary hiccup in an otherwise smoothly performing, safe operation. Nothing more fundamental, or more expensive, needs to be changed.

A patient died in an Argentine hospital because of an experimental US drug, administered to him and many fellow patients. The event was part of a clinical trial of a yet unapproved medicine eventually destined for the lucrative North American market. To many, the case was only the latest emblem of a disparity where Western nations use poorer, less scrupulous, relatively underinformed and healthcare-deprived medical testing grounds in the Second and Third World.

But the drug manufacturer was quick to point out that "the case was an aberration" and emphasized how the "supervisory and quality assurance systems all worked effectively". The system, in other words, was safe—it simply needed to be cleansed of its Bad Apples. The hospital fired the doctors involved and prosecuters were sent after them with murder charges.[4]

Saving face

In the aftermath of failure, pressure can exist to save public image. Taking out defective practitioners is always a good start to saving face. It tells people that the mishap is not a systemic problem, but just a local glitch in an otherwise smooth operation. It also shows that you are doing something about your problem. That you are taking action.

Two hard disks with classified information went missing from the Los Alamos nuclear laboratory, only to reappear under suspicious circumstances behind a photocopier a few months later. Under pressure to assure that the facility was secure and such lapses extremely uncommon, the Energy Secretary attributed the incident to "human error, a mistake". The hard drives were probably misplaced out of negligence or inattention to security procedures, officials said. The Deputy Energy Secretary added that "the vast majority are doing their jobs well at the facility, but it probably harbored 'a few bad apples' who had compromised security out of negligence".[5]

Personal responsibility and the illusion of omnipotence

Another reason for the persistence of the Bad Apple Theory is that practitioners in safety-critical domains typically assume great personal responsibility for the outcomes of their actions. Practitioners get trained and paid to carry this responsibility, and are generally quite proud of it.

But the other side of taking this responsibility is the assumption that they have the authority, the power, to match it; to live up to it. The assumption is that people can simply choose between making errors and not making them—independent of the world around them. This is an illusion of omnipotence.

The pilot of an airliner accepted a different runway with a more direct approach to the airport. The crew got in a hurry and made a messy landing that resulted in some minor damage to the aircraft. Asked why they accepted the runway, the crew cited a late arrival time and many connecting passengers on board. The investigator's reply was that real pilots are of course immune to those kinds of pressures.

The reality is that people are not immune to those pressures, and the organizations that employ them would not want them to be. People do not operate in a vacuum, where they can decide and act all-powerfully. To err or not to err is not a choice. Instead, people's work is subject to and constrained by multiple factors. Individual responsibility is not always matched by individual authority. Authority can be restricted by other people or parts in the system; by other pressures; other deficiencies.

In the VIP jet's case, it was found that there was no checklist that told pilots what to do in case of a warning light related to the pitch system. The procedure to avoid the oscillations would have been to reduce airspeed to less than 260 knots indicated. But the procedure was not in any manual. It was not available in the cockpit. And it's hardly the kind of thing you can think up on the fly.

Where the Old View Falls Short

If you are trying to explain human error, you can safely bet that you are not there yet if you have to count on individual people's negligence or complacency; if your explanation still depends on a large measure of people not motivated to try hard enough.

Something was not right with the story of the VIP jet from the start. How, really, could pilots "ignore" a light for which there was no procedure available? You cannot ignore a procedure that does not exist. Factors from the outside seriously constrained what the pilots could have possibly done. Problems existed with this particular aircraft. No procedure was available to deal with the warning light.

Whatever label is in fashion (complacency, negligence, ignorance), if a human error story is complete only by relying on Bad Apples who lack the motivation to perform better, it is probably missing the real story behind failure, or at least large parts of it.

Local rationality

The point is, people in safety-critical jobs are generally motivated to stay alive, to keep their passengers, their patients, their customers alive. They do not go out of their way to deliver overdoses; to fly into mountainsides or windshear; to amputate wrong limbs; to convict someone innocent, to lose disks. In the end, what they are doing makes sense to them at that time.

If your explanation still relies on unmotivated people, you have more work to do

It *has* to make sense, otherwise they would not be doing it. So if you want to understand human error, your job is to understand *why* it made sense to them. Because if it made sense to them, it may well make sense to other practitioners too, which means that the problem may show up again and again.

This, in human factors, is called the local rationality principle. People are doing reasonable things given their point of view and focus of attention; their knowledge of the situation; their objectives and the objectives of the larger organization they work for. In normal work that goes on in normal organizations, safety is never the only concern, or even the primary concern. Systems do not exist to be safe, they exist to make money; to render a service; provide a product. Besides safety there are multiple other objectives: pressures to produce; to not cost an organization unnecessary money; to be on time; to get results; to keep customers happy. People's sensitivity to these objectives, and their ability to juggle them in parallel with demands for safety, is one reason why they were chosen for the jobs, and why they are allowed to keep them.

> **You have to assume that nobody comes to work to do a bad job**

People do want to be bothered: their lives and livelihoods are on the line. In human factors, you have to assume that nobody comes to work to do a bad job. If people do come to work to do a bad job, then you are in the territory of sabotage, of pathological behavior. That is not what human factors is about. Human factors, its ideas, theories, principles, concepts, are about normal people doing normal work in normal organizations. And why things go right or wrong with good people trying to do a good job.

In the Los Alamos nuclear research facility, complacency was no longer a feature of a few individuals—if it ever had been. Under pressure to perform daily work in a highly cumbersome context of checking, double-checking and registering the use of sensitive materials, "complacency" (if one could still call it that) had become a feature of the entire laboratory. Scientists routinely moved classified material without witnesses or signing logs. Doing so was not a sign of malice. The practice had grown over time, bending to production pressures from which the laboratory owed its existence.[6]

If you want to understand human error, you have to assume that people were doing reasonable things given the complexities, dilemmas, trade-offs and uncertainty that surrounded them. Just finding and highlighting people's mistakes explains nothing. Saying what people did not do, or what they should have done, does not explain why they did what they did.

The point of understanding human error is to reconstruct why actions and assessments that are now controversial, made sense to people at the time. You have to push on people's mistakes until they make sense—relentlessly. You have to reconstruct, rebuild their circumstances; resituate the controversial actions and assessments in the flow of behavior of which they were part and see how they reasonably fit the world that existed around people at the time. For this, you can use all the tools and methods that the Field Guide provides.

> **You have to understand why what people did made sense to them at the time**

Notes

1. *International Herald Tribune*, June 13, 2000.
2. Lee, T. and Harrison, K. (2000). Assessing safety culture in nuclear power stations. *Safety Science, 34*, 61–97.
3. *FLIGHT International*, June 6–12, 2000.
4. *International Herald Tribune*, December 22, 2000 (p. 21).
5. *International Herald Tribune*, June 19, 2000.
6. *International Herald Tribune*, June 20, 2000.

2 The New View of Human Error

In the New View of human error:

- Human error is not a cause of failure. Human error is the effect, or symptom, of deeper trouble.
- Human error is not random. It is systematically connected to features of people's tools, tasks and operating environment.
- Human error is not the conclusion of an investigation. It is the starting point.

History is rife with investigations where the label "Human Error" was the conclusion. Paul Fitts, marking the start of aviation human factors in 1947, began to turn this around. Digging through 460 cases of "pilot error" that had been presented to him, he found that a large part consisted of pilots confusing the flap and gear handles. Typically, a pilot would land and then raise the gear instead of the flaps, causing the airplane to collapse onto the ground.

Examining the hardware in the average cockpit, Fitts found that the controls for gear and flaps were often placed next to one another. They looked the same, felt the same. Which one was on which side was not standardized across cockpits. An error trap waiting to happen, in other words. Errors (confusing the two handles) were not incomprehensible or random: they were systematic; connected clearly to features of the cockpit layout.

The years since Fitts (1947)[1] have seen an expansion of this basic idea about human error. The new view now examines not only the engineered hardware that people work with for systemic reasons behind failure, but features of people's operations and organizations as well—features that push people's trade-offs one way or another.

An airline pilot who was fired after refusing to fly during a 1996 ice storm, was awarded 10 million dollars by a jury. The pilot, who had flown for the airline for

10 years, was awarded the money in a lawsuit contending that he had been fired for turning around his turboprop plane in a storm. The pilot said he had made an attempt to fly from Dallas to Houston but returned to the airport because he thought conditions were unsafe.[2]

A hero of the jury (themselves potential passengers probably), they reasoned that this pilot could have decided to press on. But if something had happened to the aircraft as a result of icing, the investigation would probably have returned the finding of "human error", saying that the pilot knowingly continued into severe icing conditions. His trade-off must be understood against the backdrop of a turboprop crash in his company only a few years earlier—icing was blamed in that case.

An example like this confirms that:

- Systems are not basically safe. People in them have to create safety by tying together the patchwork of technologies, adapting under pressure and acting under uncertainty.
- Safety is never the only goal in systems that people operate. Multiple interacting pressures and goals are always at work. There are economic pressures; pressures that have to do with schedules, competition, customer service, public image.
- Trade-offs between safety and other goals often have to be made under uncertainty and ambiguity. Goals other than safety are easy to measure (How much fuel or time will we save? Will we get to our destination?). However, how much people borrow from safety to achieve those goals is very difficult to measure.

Trade-offs between safety and other goals enter, recognizably or not, into thousands of little and larger decisions and considerations that practitioners make every day. Will we depart or won't we? Will we push on or won't we? Will we operate or won't we? Will we go to open surgery or won't we? Will we accept the direct or won't we? Will we accept this indication or alarm as indication of trouble or won't we?

Systems are not basically safe. People create safety while negotiating multiple system goals

These trade-offs need to be made under much uncertainty and often under time pressure. In the new view on human error:

- People are vital to creating safety. They are the only ones who can negotiate between safety and other pressures in actual operating conditions;

- Human errors do not come unexpectedly. They are the other side of human expertise—the human ability to conduct these negotiations while faced with ambiguous evidence and uncertain outcomes.

Research Insights Behind the New View

The new view of human error centers around a number of insights. These have grown as research has learned more and more about complex systems over the past decades:

- **Sources of error are structural, not personal.** If you want to understand human error, you have to dig into the system in which people work. You have to stop looking for people's personal shortcomings;
- **Errors and accidents are only remotely related.** Accidents emerge from the system's complexity, not from its apparent simplicity. That is, accidents do not just result from a human error, or a procedural "violation". It takes many factors, all necessary and only jointly sufficient, to push a system over the edge of failure;
- **Accidents are not the result of a breakdown** of otherwise well-functioning processes. You think your system is basically safe and that accidents can only happen if somebody does something really stupid or dangerous. Instead, the research is showing us how accidents are actually structural by-products of a system's normal functioning.

These three insights stand in sharp contrast to what drives the Old View, or Bad Apple Theory. It, after all, assumes that mistakes are a result of personal shortcomings. For example, people did not try hard enough. They should have looked a bit better, or concentrated a bit more. The Old View also sees an obvious link between people's mistakes, or people's rule-breaking, and accidents. Accidents would not happen if people just followed the procedures or regulations. Indeed, the Old View sees accidents mostly as the result of people not performing well in an otherwise safe system.

Failure as the by-product of normal work

What is striking about many mishaps is that people were doing exactly the sorts of things they would usually be doing—the things that usually lead to success and safety. People are doing what makes sense given the situational indications, operational pressures and organizational norms existing at the time. Accidents are seldom preceded by bizarre behavior.

If this is the most profound lesson you and your organization can learn from a mishap, it is also the most frightening. It means acknowledging that failures are baked into the very nature of your work and organization; that they are symptoms of deeper trouble or by-products of systemic brittleness in the way you do your business. It means having to acknowledge that mishaps are the result of everyday influences on everyday decision making, not isolated cases of erratic individuals behaving unrepresentatively. It means having to find out why what people did back there and then actually made sense given the organization and operation that surrounded them.

Investigations and the New View

In the New View, investigations are driven by one unifying principle: human errors are symptoms of deeper trouble. This means that human error is the starting point of an investigation. You are interested in what the error points to. What are the sources of people's difficulties? You have to target what lies behind the error—the organizational trade-offs pushed down into individual operating units; the effect of new technology; the complexity buried in the circumstances surrounding human performance; the nature of the mental work that went on in difficult situations; the way in which people coordinated or communicated to get their jobs done; the uncertainty of the evidence around them.

Human error is a symptom of trouble deeper in the system

Why are investigations in the new view interested in these things? Because this is where the action is. If you want to learn anything of value about the systems we operate, you must look at human errors as:

- A window on a problem that every practitioner in the system might have;
- A marker in the system's everyday behavior, and an opportunity to learn more about organizational, operational and technological features that create error potential.

Recommendations in the new view:

- Are hardly ever about individual practitioners, because their errors are a symptom of systemic problems that everyone may be vulnerable to;

- Do not rely on tighter procedures because humans need the discretion to deal with complex and dynamic circumstances for which pre-specified guidance is badly suited;
- Do not get trapped in promises of new technology. Although it may remove a particular error potential, new technology will likely present new complexities and error traps.
- Try to address the kind of systemic trouble that has its source in organizational decisions, operational conditions or technological features.

The New View and Progress on Safety

The New View of human error does not claim that people are perfect. Goals aren't met; selections are wrongly made; situations get misassessed. In hindsight it is easy to see all of that. People inside the situations themselves may even see that.

But the New View avoids judging people for this. It wants to go beyond saying what people should have noticed or could have done. Instead, the New View seeks to explain "why". It wants to understand why people made the assessments or decisions they made—why these assessments or decisions would have made sense from the point of view inside the situation. When you see people's situation from the inside, as much like these people did themselves as you can reconstruct, you may begin to see that they were trying to make the best of their circumstances, under the uncertainty and ambiguity surrounding them. When viewed from inside the situation, their behavior probably made sense—it was systematically connected to features of their tools, tasks and environment.

The Field Guide helps you understand human error according to the New View. It intends to help you identify how people's assessments and actions could have made sense to them, according to their circumstances at the time. It will help you avoid the traps of judging people for mistakes, and instead see those mistakes as interesting windows on the functioning of the entire system of which people were part. It will help you build a vocabulary and set of techniques for applying the New View. So you, in turn, can help your organization learn something valuable from failure, instead of just saying "human error".

The point is not to see where people went wrong, but why what they did made sense

Notes

1 Fitts, P.M. and Jones, R.E. (1947). Analysis of factors contributing to 460 "pilot error" experiences in operating aircraft controls. *Memorandum Report TSEAA-694-12*, Aero Medical Laboratory, Air Material Command, Wright-Patterson Air Force Base, Dayton, Ohio, 1 July 1947.
2 *International Herald Tribune*, January 15, 2000.

3 The Hindsight Bias

Have you ever caught yourself asking, "How could they not have noticed?" or "How could they not have known?" Then you were reacting to failure. And to understand failure, you first have to understand your reactions to failure.

We all react to failure. In fact, our reactions to failure often make that we see human error as the cause of a mishap; they promote the Bad Apple Theory. Failure, or people doing things with the potential for failure, is generally not something we expect to see. It surprises us; it does not fit our assumptions about the system we use or the organization we work in. It goes against our beliefs. As a result, we try to reduce that surprise—we react to failure.

> *To understand failure, you must first understand your reactions to failure*

A Navy submarine crashed into a Japanese fishing vessel near Hawaii, sinking it and killing nine Japanese men and boys. The submarine, on a tour to show civilians its capabilities, was demonstrating an "emergency blow"—a rapid re-surfacing. Time had been running short and the crew, crowded in the submarine's control room with sixteen visitors, conducted a hurried periscope check to scan the ocean surface. Critical sonar equipment onboard the submarine was inoperative at the time.

The commander's superior, an Admiral, expressed shock over the accident. He was puzzled, since the waters off Hawaii are among the easiest areas in the world to navigate. According to the admiral, the commander should not have felt any pressure to return on schedule. At one of the hearings after the accident, the Admiral looked at the commander in the courtroom and said "I'd like to go over there and punch him for not taking more time". As the Admiral saw it, the commander alone was to blame for the accident—civilians onboard had nothing to do with it, and neither had inoperative sonar equipment.[1]

Reactions to failure, such as in the example above, share the following features:

- **Retrospective**. Reactions arise from our ability to look back on a sequence of events, of which we know the outcome;
- **Counterfactual**. They lay out in detail what people could or should have done to prevent the mishap;

- **Judgmental.** They judge people (e.g., not taking enough time, not paying enough attention, not being sufficiently motivated) for supposed personal shortcomings;
- **Proximal.** They focus on those people who were closest in time and space to the mishap, or to potentially preventing it.

Reactions to failure interfere with your understanding of failure. The more you react, the less you understand. When you look closely at findings and conclusions about human error, you can see that they are often driven by reactions to failure, and written in their language.

The next few chapters are about these reactions, how to recognize them, how to avoid them and how not to talk or write in their language. The first of these chapters, this one, is about the hind-sight bias. This is one of the most powerful biases documented in psychology. It is fundamental to the misunderstandings that you introduce to your analysis of past performance. Understanding the hindsight bias, and knowing its effects, is crucial for moving beyond your reactions to failure. Controlling the hindsight bias is critical for understanding human error.

The more you react to failure, the less you will understand

Reactions to Failure are Retrospective

Investigations aim to explain a part of the past. Yet they are conducted in the present, and thus inevitably influenced by it. As investigator, you are likely to know:

- The outcome of a sequence of events you are investigating;
- Which cues and indications were critical in the light of the outcome—what were the signs of danger?
- Which actions would have prevented the outcome.

A highly automated airliner crashed on a golf course short of the runway at an airport in India. During the final approach, the aircraft's automation had been in "open descent mode", which manages airspeed by pitching the nose up or down, rather than through engine power. When they ended up too low on the approach, the crew could not recover in time. In hindsight, the manufacturer of the aircraft commented that "the crew should have known they were in open descent mode". Once outside observers learned its importance, the question became how the crew could have missed or misunderstood such a critical piece of information.

You Probably Know More than the People Involved

One of the safest bets you can make as an investigator or outside observer is that you know more about the incident or accident than the people who were caught up in it—thanks to hindsight:

- Hindsight means being able to look back, from the outside, on a sequence of events that led to an outcome you already know about;
- Hindsight gives you almost unlimited access to the true nature of the situation that surrounded people at the time (where they actually were versus where they thought they were; what state their system was in versus what they thought it was in);
- Hindsight allows you to pinpoint what people missed and shouldn't have missed; what they didn't do but should have done.

The Chairman of the investigation into the Clapham Junction railway accident in Britain wrote, "There is almost no human action or decision that cannot be made to look flawed and less sensible in the misleading light of hindsight. It is essential that the critic should keep himself constantly aware of that fact."[2]

Hindsight biases your investigation towards items that you *now* know were important (e.g. "open descent mode"). As a result, you may assess people's decisions and actions mainly in the light of their failure to pick up this critical piece of data. It artificially narrows your examination of the evidence and potentially misses alternative or wider explanations of people's behavior.

The effect of knowing an outcome of a sequence of events is huge. It has an enormous impact on your ability to objectively look back on a piece of performance. Actually, you no longer can.

Consider the Greek mythological figure of Oedipus. He went about merrily, getting invited to make love to a woman named Jocasta. Which he did. It was only after the intimate encounter that he learned, from a messenger, that Jocasta was actually his mother. What do you think the difference is between Oedipus' memory of the brief affair —before and after he got the message? Knowing the outcome, he would no longer have been able to look back objectively on his behavior. In fact, he would probably go around asking himself how he could not have noticed, where he failed to double-check, what he had missed, what he had misjudged. He would, in other words, discover all kinds of "errors". Without hindsight, without the messenger, these "errors" would never have existed.

The Illusion of Cause-Consequence Equivalence

The reason why hindsight destroys our ability to look objectively at past performance has to do with "cause-consequence equivalence". This says that a bad outcome can only have been preceded by a bad process. Indeed, we assume that really bad consequences can only be the result of really bad causes. Faced with a disastrous outcome, or the potential for one, we assume that the acts leading up to it must have been equally monstrous. So once we know an outcome is bad, we can no longer look objectively at the process that led up to it.

We assume that bad outcome = bad process

But this automatic response is very problematic in complex worlds. Here even bad processes often lead to good outcomes. And good processes can lead to bad outcomes. Processes may be "bad" in the retrospecitve sense that they departed from routines you now know to have been applicable. But this does not necessarily lead to failure. Given their variability and complexity, these worlds typically offer an envelope of options and pathways to safe outcomes. There is more than one way to success. Think of a rushed approach in an aircraft that becomes stabilized at the right time and leads to a safe landing. This can actually serve as a marker of expertise, where people successfully juggle pressures for production with an ability to stay safe. The opposite goes too. Good processes, where people double-check and communicate and stick to procedures, can lead to disastrous outcomes if only you surround them with the right unusual circumstances.

Hindsight Determines How You Know History

With knowledge of outcome, our understanding of events changes dramatically. In fact, hindsight is responsible for the way we know history. It determines how we know history (see Figure 3.1).

Hindsight causes us to oversimplify history, relative to how people understood events at the time they were happening. This oversimplification becomes apparent in how:

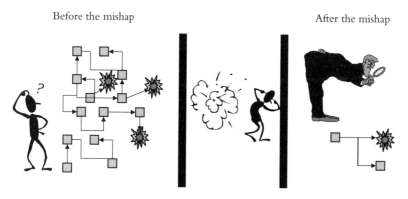

Figure 3.1 Hindsight changes how we look at past decision making. It turns real, convoluted complexity into a simple, linear story; a binary decision to err or not to err. Idea for image by Richard Cook.

- We think that a sequence of events inevitably led to an outcome. We underestimate the uncertainty people faced at the time, or do not understand how very unlikely the actual outcome would have seemed. Had we seen the same situation from the inside, we would understand that the outcome (that we now know about) was once an infinitesimally small probability; one among many other possible outcomes;
- We see a sequence of events as linear, leading nicely and uninterruptedly to the outcome we now know about. Had we seen the same situation from the inside, we would have recognized the possible confusion of multiple possible pathways and many zigs and zags surrounding people;
- We oversimplify causality. When we are able to trace a sequence of events backwards (which is the opposite from how people experienced it at the time), we easily couple "effects" to preceding "causes" (and *only* those causes) without realizing that causal couplings are much more difficult to sort out when in the middle of things.

Inside the tunnel

The outcome of a sequence of events is the starting point of your work as investigator. Otherwise you wouldn't actually be there. This puts you at a remarkable disadvantage when it comes to understanding the point of view of the people you're investigating. Tracing back from the outcome, you will come across joints where people had opportunities to "zig" instead of "zag"; where they could have directed the events away from failure. As investigator you come out on the other end of the sequence of events wondering how people could have missed those opportunities to steer away from failure.

Look at Figure 3.1. You see an unfolding sequence of events there. It has the shape of a tunnel which is meandering its way to an outcome. The figure shows two different perspectives on the pathway to failure:

- **The perspective from the outside and hindsight** (typically your perspective). From here you can oversee the entire sequence of events—the triggering conditions, its various twists and turns, the outcome, and the true nature of circumstances surrounding the route to trouble.
- **The perspective from the inside of the tunnel**. This is the point of view of people in the unfolding situation. To them, the outcome was not known (or they would have done something else). They contributed to the direction of the sequence of events on the basis of what they saw on the *inside* of the unfolding situation. To understand human error, you need to attain this perspective.

The Field Guide invites you to go inside the tunnel of Figure 3.2. It will help you understand an evolving situation from the point of view of the people inside of it, and see why their assessments and actions made sense at the time.

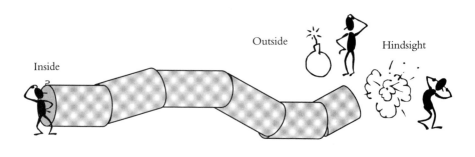

Figure 3.2   Different perspectives on a sequence of events: Looking from the outside and hindsight you have knowledge of the outcome and dangers involved. From the inside, you may have neither.

Hindsight Is Everywhere

Hindsight is baked deeply into the language of accident stories we tell one another. Take a common problem today—people losing track of what mode their automated systems are operating in. This happens in cockpits, operating rooms, process control plants and many other workplaces. In hindsight, when you know how things developed and turned out, this problem is often called "losing mode awareness". Or, more broadly, "loss of situation awareness". What are you really saying? Not much, other than that you, in your omniscient position after-the-fact, know more about the situation than the people who were in the middle of it. Indeed, loss of situation awareness is the difference between:

- what you *now* know the situation actually was like;
- what people understood it to be at the time.

It is easy to show that people at another time and place did not know what you know today ("they should have known they were in open descent mode"). But it is not an explanation of their behavior. You must guard yourself against mixing your reality with the reality of the people you are investigating. Those people did not know there was going to be a negative outcome, or they would have done something else. It is impossible for people to assess their decisions or incoming data in light of an outcome they do not yet know about.

> *"Loss of situation awareness" is the difference between what you now know, and what other people knew back then*

Even in less obvious remarks, hindsight can reign surpeme:

> Throughout history poor decision making has led to undesired outcomes. In the first days of World War I, for example, the military commanders of several countries made decisions that led to disastrous consequences. In the face of overwhelming evidence that their initial plans needed to be revised in the light of battlefield conditions, they refused, decisions that ultimately led to their countries' defeat.[3]

While such statements seem sensible at first, they actually offer very little leverage for making progress on people's and systems' performance. Human error (poor decision making) is the cause of trouble, which people seem to be willingly wreaking upon their country (they "refused" to change plans).

It is hindsight that helps make the decisions poor, hindsight that allows you to see which evidence was "overwhelming" in its suggestion that people should be doing things differently, and hindsight that neatly, linearly couples people's decisions to disastrous outcomes, downplaying or ignoring the confounding effects of context, surprise, history and institutional constraints that operate on decision makers of this kind. Unfortunately, all the interesting questions about these things (why it makes sense for military commanders to continue with original plans, and why others will probably do the same thing again) go unasked.

Hindsight means the Old View. The Old View means hindsight. This will forever keep you from really understanding human error.

Notes

1   *International Herald Tribune*, March 14, 2001.
2   Hidden, A. (1989). *Clapham Junction Accident Investigation Report*, p. 147.
3   Strauch, B. (2002). *Investigating human error: Incidents, accidents and complex systems.* Aldershot, UK: Ashgate Publishing Co., p. 197.

4 Put Data in Context

You can avoid the hindsight bias by putting yourself in the shoes of the people whose behavior you are trying to understand.

- How did the world look to them at the time?
- How did the situation unfold around them; what cues did they get when?
- What goals were they likely pursuing at that time (not knowing the outcome you now know about)?

Remember, you want to understand why it made sense for these people to do what they did. That means that you need to put the data that you have on what they did, back into *their* context. You need to put people's behavior back into the situation that produced and accompanied that behavior.

But this is not easy. In fact, it is always tempting to go for a context that actually lies *outside* the mishap sequence. Taking behavior out of context, and giving it meaning from the outside, is common in efforts to understand human error. Hindsight produces various ways in which behavioral data gets taken out of context:

- **Micro-matching** data with a world you now know to be true and finding a mismatch;
- **Cherry-picking** selected bits that prove a condition you have identified only in hindsight;
- Presenting a **shopping bag** full of cues and indications that pointed to the real nature of the situation, and wondering how people could possibly have missed all that evidence.

Out of Context I: Micro-Matching

One of the most popular ways you can assess performance after-the-fact is to hold it up against a world you *now* know to be true. There are various after-the-fact-worlds that you can bring to life:

- A procedure or collection of rules: People's behavior was not in accordance with standard operating procedures that were found to be applicable for that situation afterward;
- A set of cues: People missed cues or data that turned out to be critical for understanding the true nature of the situation;
- Standards of good practice: People's behavior fall short of standards of good practice in the particular industry.

The problem is that these after-the-fact-worlds may have very little in common with the actual world that produced the behavior under investigation. They contrast people's behavior against the investigator's reality, not the reality that surrounded the behavior in question. Thus, micro-matching fragments of behavior with these various standards explains nothing—it only judges.

Imposing procedures onto history

First, individual fragments of behavior are frequently compared with procedures or regulations, which can be found to have been applicable in hindsight. Compared with such written guidance, actual performance is often found wanting; it does not live up to procedures or regulations.

Take the automated airliner that started to turn towards mountains because of a computer-database anomaly. The aircraft ended up crashing in the mountains. The accident report explains that one of the pilots executed a computer entry without having verified that it was the correct selection, and without having first obtained approval of the other pilot, contrary to the airline's procedures.[1]

Investigations invest considerably in organizational archeology to construct the regulatory or procedural framework in which operations took (or should have taken) place. In hindsight, you can easily expose inconsistencies between rules and actual behavior. Your starting point is a fragment of behavior, and you have the luxury of time and resources to excavate organizational records and regulations to find rules with which the fragment did not match (Figure 4.1).

But what have you shown? You have only pointed out that there was a mismatch between a fragment of human performance and existing guidance that you uncovered or highlighted after-the-fact.

This is not very informative. Showing that there was a mismatch between procedure and practice sheds little light on the *why* of the behavior in question. And, for that matter, it sheds little light on the why of this particular mishap.

Put Data in Context 31

Figure 4.1 Micromatching can mean that you take performance fragments from the stream of events and hold them up against rules or procedures that you deem applicable in hindsight. You don't explain anything by doing this.

Mismatches between procedure and practice are not unique ingredients of accident sequences. They are often a feature of daily operational life (which is where an interesting bit in your understanding of human error starts).

Imposing available data onto history

Another way to construct the world against which to evaluate individual performance fragments, is to turn to data in the situation that was not noticed but that, in hindsight, turned out to be critical.

Continue with the automated aircraft. What should the crew have seen in order to notice the turn? They had plenty of indications, according to the manufacturer of their aircraft:
  "Indications that the airplane was in a left turn would have included the following: the EHSI (Electronic Horizontal Situation Indicator) Map Display (if selected) with a curved path leading away from the intended direction of flight; the EHSI VOR display, with the CDI (Course Deviation Indicator) displaced to the right, indicating the airplane was left of the direct Cali VOR course, the EaDI indicating approximately 16 degrees of bank, and all heading indicators moving to the right. Additionally the crew may have tuned Rozo in the ADF and may have had bearing pointer information to Rozo NDB on the RMDI".[2]

This is a standard response after mishaps: point to the data that would have revealed the true nature of the situation. But knowledge of the "critical" data comes only with the privilege of hindsight. If such critical data can be shown to have been physically available, it is automatically assumed that it should have been picked up by the operators in the situation. Pointing out, however, that it should have been does not explain why it was perhaps not, or why it was interpreted differently back then. There is a difference between:

- **Data availability**: what can be shown to have been physically available somewhere in the situation;
- **Data observability**: what would have been observable given the features of the interface and the multiple interleaving tasks, goals, interests, knowledge and even culture of the people looking at it.

The mystery for understanding human error is not why people could have been so unmotivated or unwise not to pick up the things that you can decide were critical in hindsight. The mystery—and your job—is to find out what *was* important to them, and why.

Imposing other standards onto history

Third, there are a number of other standards, especially for performance fragments that do not easily match procedural guidance. Or for which it is more difficult to point out data that existed in the world and that people "should" have picked up.

This is often the case when a controversial fragment knows no clear pre-ordained guidance but relies on local, situated judgment. Take, for example, a decision to accept a runway change, or to continue flying into bad weather. For these cases there are always "standards of good practice" which are based on convention and putatively practiced across an entire industry. One such standard in aviation is "good airmanship", which, if nothing else can, will cover the variance in behavior that had not yet been accounted for. Other industries do this too:

Cases for medical negligence can often be made only by contrasting actual physician performance against standards of proper care or good practice. Rigid procedures generally cannot live up to the complexity of the work and the ambiguous, ill-defined situations in which it needs to be carried out. Consequently, it cannot easily be claimed that this or that checklist should have been followed in this or that situation.

But which standards of proper care do you invoke to contrast actual behavior against? This is largely arbitrary, and driven by hindsight. After wrong-site surgery, for example, the standard of good care that gets invoked is that physicians have to make sure that the correct limb is amputated or operated upon. Finding vague or broad standards in hindsight does nothing to elucidate the actual circumstances and systemic vulnerabilities which in the end allowed wrong-site surgery to take place.

By referring to procedures, physically available data or standards of good practice, you can micro-match controversial fragments of behavior with standards that seem applicable from your after-the-fact position.

You construct a referent world from outside the accident sequence, based on data you now have access to, based on facts you now know to be true. The problem is that these after-the-fact-worlds may have very little relevance to the circumstances of the accident sequence. They do not explain the observed behavior. You have substituted your own world for the one that surrounded the people in question.

Out of Context II: Cherry-Picking

The second way in which you can take data out of context, in which you give them meaning from the outside, is by grouping and labeling behavior fragments that, in hindsight, appear to represent a common condition.

Consider this example, where diverse fragments of behavior are lumped together to build a case for haste as explanation of the bad decisions taken by the crew. The fragments are actually not temporally co-located. They are spread out over a considerable time, but that does not matter. According to the investigation they point to a common condition.

"Investigators were able to identify a series of errors that initiated with the flightcrew's acceptance of the controller's offer to land on runway 19 ... The CVR indicates that the decision to accept the offer to land on runway 19 was made jointly by the captain and the first officer in a 4-second exchange that began at 2136:38. The captain asked: 'would you like to shoot the one nine straight in?' The first officer responded, 'Yeah, we'll have to scramble to get down. We can do it.' This interchange followed an earlier discussion in which the captain indicated to the first officer his desire to hurry the arrival into Cali, following the delay on departure from Miami, in an apparent effort to minimize the effect of the delay on the flight attendants' rest requirements. For example, at 2126:01, he asked the first officer to 'keep the speed up in the descent' ... The evidence of the hurried nature of the tasks performed and the inadequate review of critical information between the time of the flightcrew's acceptance of the offer to land on runway 19 and

the flight's crossing the initial approach fix, ULQ, indicates that insufficient time was available to fully or effectively carry out these actions. Consequently, several necessary steps were performed improperly or not at all".[3]

As one result of the runway change and self-imposed workload the flight crew also "lacks situation awareness"—an argument that is also constructed by grouping voice utterance fragments from here and there:

"... from the beginning of their attempt to land on runway 19, the crew exhibited a lack of awareness .... The first officer asked 'where are we?', followed by 'so you want a left turn back to ULQ?'. The captain replied, 'hell no, let's press on to ... and the first officer stated 'well, press on to where though?' .... Deficient situation awareness is also evident from the captain's interaction with the Cali air traffic controller".[4]

It is easy to pick through the evidence of an accident sequence and look for fragments that all seem to point to a common condition. The investigator treats the voice record as if it were a public quarry to select stones from, and the accident explanation the building he needs to construct from those stones. Among investigators this practice is sometimes called "cherry-picking"—selecting those bits that help their *a priori* argument. The problems associated with cherry-picking are many:

- You probably miss all kinds of details that are relevant to explaining the behavior in question;
- Each cherry, each fragment, is meaningless outside the context that produced it. Each of the bits that gets lumped together with other "similar" ones actually has its own story, its own background, its own context and its own reasons for being. When it was produced it may have had nothing to do with the other fragments it is now grouped with. The similarity is entirely in the eye of the retrospective beholder.
- Much performance, much behavior, takes place *in between* the fragments that the investigator selects to build his case. These intermediary episodes contain changes and evolutions in perceptions and assessments that separate the excised fragments not only in time, but also in meaning.

Thus, the condition that binds similar performance fragments together has little to do with the circumstances that brought each of the fragments forth; it is not a feature of those circumstances. It is an artifact of you as outside observer. The danger is that you come up with a theory that guides the search for evidence about itself. This leaves your understanding of human error with tautologies, not explanations.

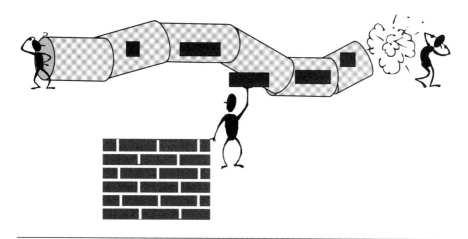

Figure 4.2 Cherry-picking means taking fragments from all over the record and constructing a story with them that exists only in your hindsight. In reality, those pieces may have had nothing to do with each other.

Out of Context III: The Shopping Bag

With the benefit of hindsight, it is so easy to sweep together all the evidence that people should have seen. If they had, they would have recognized the situation for what we now know it turned out to be. Arguments here sometimes go beyond mixing up data availability and observability. Hindsight has a way of easily organizing all the evidence pointing to the (bad) outcome. But that doesn't mean the evidence presented itself that way to people at the time.

Airplanes sometimes run off the end of the runway after landing. This may happen because the landing speed is too high, or the weather is bad, with a lot of tailwind, or the runway wet and slippery. I remember an accident where a combination of these things played a role, and I was asked to explain how a professional crew could have missed all the evidence that pointed to a deteriorating weather situation at the arrival airport. The wind had been shifting around, there was a thunderstorm moving about the area, the visibility was getting worse, and it was raining. It was as if my head was shoved into a shopping bag full of epiphanies, and I was asked "look, how could they not have seen all that evidence and concluded that attempting a landing was a bad idea?"

The question was misleading and driven entirely by hindsight. Only hindsight made my questioners able to chuck all the cues and indications about bad weather together in one bag. But this is not how these cues and indications revealed themselves to the pilots at the time! In fact, there was a lot of evidence, compelling, strong and early on, that the weather was going to be just fine. Cues and indications that the weather was deteriorating then came dripping in, one by one. They contradicted one another; they were not very compelling.

If you want to understand why the pilots did what they did, I replied, you have to reconstruct the situation as it unfolded around them. When did which cues come in? What did they likely mean given the context in which they appeared? What would that have meant for how the understanding of the pilots developed over time?

By sweeping cues and indications about an unfolding situation together and presenting them as one big glob of overwhelming evidence, you import our own reality into the situation that surrounded other people at another time. That way you will never understand why it could have made sense for them to do what they did. You will simply be left wondering how they could have missed what seems to add up to such an obvious picture to you now.

That picture on the inside of the shopping bag is obvious only if you know the outcome. It is outcome that offers you the shopping bag, that binds the cues and indications together.

The people whose behavior you are trying to understand did not know the outcome (or they would have done something else). In order to understand human error, you have to put yourself in their shoes. Imagine that you don't know the outcome. Try to re-construct which cues came when, which indications may have con-tradicted them. Envisage what this unfolding trickle of cues and indications could have meant to those other people, given their likely under-standing of the situation at the time (remember, you are trying not to know the outcome). Try to understand how their understanding of the situation was not static or complete, as yours in the shopping bag is, but rather incomplete, unfolding and uncertain.

Step Into the Past and Look Around

Taking data out of context, either by:

- micro-matching them with a world you now know to be true;
- lumping selected bits together under one condition identified in hindsight, or;
- chucking cues and indications together in a shopping bag only to wonder how people could have missed all of that,

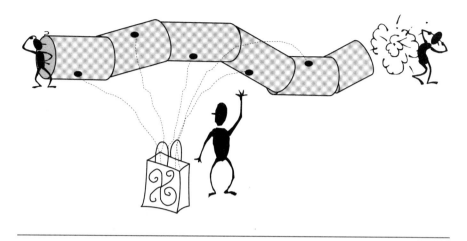

Figure 4.3 It is easy to gather cues and indications from a sequence of events and lob them together as in a shopping bag. This is not, however, how people inside the unfolding situation saw those cues presented to them.

robs data of their original meaning. And these data out of context are simultaneously given a new meaning—imposed from the outside and from hindsight. You impose this new meaning when you look at the data in a context you *now* know to be true. Or you impose meaning by tagging an outside label on a loose collection of seemingly similar fragments. Or you impose the real nature of the situation by sweeping together all the evidence you can retrospectively find for it.

To understand the actual meaning that data had at the time and place it was produced, you need to step into the past yourself. When left or relocated in the context that produced and surrounded it, human behavior is inherently meaningful.

Historian Barbara Tuchman put it this way: "Every scripture is entitled to be read in the light of the circumstances that brought it forth. To understand the choices open to people of another time, one must limit oneself to what they knew; see the past in its own clothes, as it were, not in ours."[5]

If you really want to understand human error, that is, if you really want to understand why people did what they did, nothing is as enlightening as putting

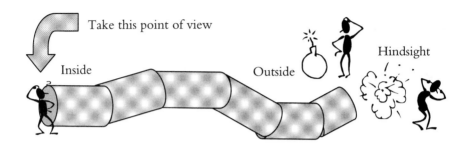

Figure 4.4 See the unfolding world from the point of view of people inside the situation—not from the outside or from hindsight

yourself in their shoes and looking around at the world as it must have looked to them.

Of course, you can never be entirely sure that your re-creation of their world matches how it must have looked to them. But at least it brings the two perspectives closer. At least it gives you a way to try to limit the effects of hindsight. Do not substitute your reality for the one that surrounded them at the time, for it will destroy your ability to make sense of other people's behavior.

Notes

1 Aeronautica Civil (1996). *Aircraft Accident Report: Controlled flight into terrain American Airlines flight 965, Boeing 757-223, N851AA near Cali, Colombia, December 20, 1995*. Santafe de Bogota, Colombia: Aeronautica Civil Unidad Administrativa Especial.
2 Boeing submission to the American Airlines Flight 965 Accident Investigation Board (1996). Seattle, WA: Boeing.
3 Aeronautica Civil, op. cit., p. 29.
4 Aeronautica Civil, op. cit., pp. 33–34.
5 Tuchman, B. (1981). *Practicing history: Selected essays*. New York: Norton, p. 75.

5 "They Should Have …"

It is so easy to say what people should have done. Or to say what they shouldn't have done. With the benefit of hindsight, you can easily see what people could have done to not have an incident or accident happen. You may even think that this explains the mishap; that you can understand human error by saying what people should have done or did not do. But this is an illusion. As soon as you say any of the following:

- "they shouldn't have…"
- "they could have…"
- "they didn't…"
- "they failed to…"
- "if only they had…!"

you are quickly gliding away from the possibility of understanding human error. We can call these phrases *counterfactuals*. They are literally saying something "counter the facts". They make you spend your time talking about a reality that did not happen (but if it had happened, the mishap would not have happened).

It is easy to fall into the fallacy that talking about things that did not happen will get you closer to understanding human error. It won't. Why would you want to waste effort on laying out what did not happen? Your charter in understanding human error is to find out why things happened the way they did. What you think should have happened or could have happened instead is, for all intents and purposes, irrelevant. Remember, you want to know why what people did made sense to them at the time. One of the first things you need to do, then, is stop talking in the language of counterfactuals. This chapter talks more about counterfactuals, and the next chapter suggests how you can avoid them.

> *What (you think) should've happened cannot explain people's behavior*

Finding Out What Could Have Prevented the Mishap

The outcome of a mishap is the starting point for your efforts to understand human error. Otherwise you probably wouldn't even bother. Knowing the outcome, and using that as your foundation, puts you at a remarkable disadvantage when it comes to seeing the point of view of the people whose errors you are trying to understand. Tracing back from the outcome, you will come across joints where people had opportunities to "zig" instead of "zag"; where they could have directed events away from failure. You come out on the other end of a sequence of events wondering how people could have missed those opportunities to steer away from failure.

Accident reports are generally full of counterfactuals that describe in fine detail the pathways and options that the people in question did not take. For example, "The airplane could have overcome the windshear encounter if the pitch attitude of 15 degrees nose-up had been maintained, the thrust had been set to 1.93 EPR (Engine Pressure Ratio) and the landing gear had been retracted on schedule."[1]

Counterfactuals prove what could have happened if certain minute and often utopian conditions had been met. Counterfactual reasoning may thus be a fruitful exercise when recommending countermeasures against such failures in the future.

But when it comes to explaining behavior, counterfactuals contribute little. Stressing what was not done (but if it had been done, the accident wouldn't have happened) explains nothing about what actually happened, or why. Counterfactuals are not opportunities missed by the people you are investigating. Counterfactuals are products of your hindsight. Hindsight allows you to transform an uncertain and complex sequence of events into a simple, linear series of obvious options. By stating counterfactuals, you are probably oversimplifying the decision problems faced by people at the time (Figure 5.1).

Explaining Failure by Seeking Failure

The effect of wandering backwards through a sequence of events is profound. You begin with the outcome failure. And to explain that failure, your first reflex is often to seek other failures. Where did people go wrong? What did they miss? To explain why an outcome failure occurred, you look for errors,

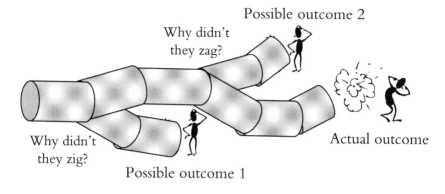

Figure 5.1 Counterfactuals: Going back through a sequence, you wonder why people missed opportunities to direct events away from the eventual outcome. This, however, does not explain failure.

for incorrect actions, flawed analyses, inaccurate perceptions. The compound picture you often get is one incremental slide into bad judgment: failure after failure after breakdown after failure. A chain of errors, a chain of misassessments, miscommunications, wrong decisions.

But you are walking backward. *With* knowledge of outcome. Leisurely sampling the various choice moments you think people had. The forks in the road stand out so clearly to you as the various choice moments converge around you. You can stop and study them, you can take your time, and you can mentally play with the idea of doing one versus the other thing here or there. No pressure, no uncertainty. After all, you know the outcome. That's where you started. And that's why you are there.

Now imagine doing it the other way around. Without knowing the outcome. You are inside the tunnel. And facing forward. And being pushed ahead by unfolding events. Time is ticking along. Now the forks are shrouded in uncertainty and the complexity of many possible options and demands; there are many more prongs and possible traps than you ever saw when you were traveling backward. And they are all surrounded by time constraints and other pressures. The decisions, judgments and perceptions that you were condemning just a few moments ago are now just that: decisions, judgments and perceptions—if they are that at all. And they must have made sense to the people making them, given their goals, knowledge and attention at the time. Otherwise these people would have done something else.

Don't Use the Word "Failure"

The very use of the word "failure" (for example: "the crew failed to recognize a mode change") indicates that you are still on the outside of the tunnel, looking back and looking down. You are handing down a judgment from outside the situation. You are not providing an explanation from people's point of view within.

The word failure implies an alternative pathway, one which the people in question did not take (for example, recognizing the mode change). Laying out this pathway is counterfactual, as explained above. By saying that people "failed" to take this pathway—in hindsight the right one—you judge their behavior according to a standard you can impose only with your broader knowledge of the mishap, its outcome and the circumstances surrounding it. You have not explained a thing yet. You have not shed light on how things looked on the inside of the situation; why people did what they did given their circumstances.

The literature on medical error describes how cases of death due to negligence may be a result of a judgment failure in the diagnostic or therapeutic process. Examples include a misdiagnosis in spite of adequate data, failure to select appropriate diagnostic tests or therapeutic procedures, and delay in diagnosis or treatment.[2]

Although they look like explanations of error, they are in fact judgments that carry no explanation at all. For example, the "misdiagnosis in spite of adequate data" was once (before hindsight) a reasonable diagnosis based on the data that was available, and seemed critical or relevant—otherwise it would not have been made by the physician in question. Calling it a misdiagnosis is an unconstructive, retrospective judgment that misses the reasons behind the actual diagnosis.

Remember the charter of the New View: You want to understand why it made sense for people to do what they did. Because if it made sense to them, it will probably make sense to other practitioners too. And trouble may repeat itself.

The word "failure" is still popular in the probable cause statements of some investigative bodies. Take, for example, the first three probable causes of the windshear accident that was referred to earlier (under Counterfactuals):

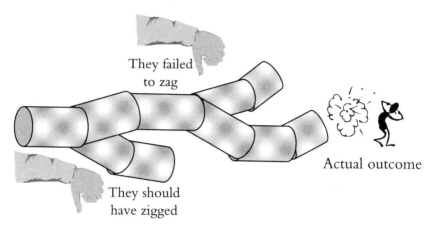

Figure 5.2 Judgmental: by claiming that people should have done something they didn't, or failed to do something they should have, you do not explain their actual behavior.

> The board determines that the probable causes of the accident were: 1) the flightcrew's decision to continue an approach into severe convective activity that was conducive to a microburst; 2) the flightcrew's failure to recognize a windshear situation in a timely manner; 3) the flightcrew's failure to establish and maintain the proper airplane attitude and thrust setting necessary to escape the windshear.

Saying what people failed to do, or implying what they could or should have done to prevent the mishap, has no role in understanding human error. For instance, failing "to recognize a windshear situation" does not begin to explain how and why the crew interpreted their situation the way they did. And it does not begin to lay out how we could help other crews get caught in the same situation.

Using this kind of language prevents you from understanding error. You simply keep occupying your judgmental perch, looking back down onto a sequence of events whose circumstances and outcome are now clear to you. To understand error, you have to climb down, take the view from the inside of the tunnel and stop saying what people failed to do or should have done. Read the next chapter to find out how.

Notes

1. National Transportation Safety Board (1995). Aircraft Accident Report: Flight into terrain during missed approach USAir flight 1016, DC-9-31, N954VJ, Charlotte, NC, July 2, 1994. Washington, DC: NTSB, pp. 119–120.
2. Bogner, M.S. (ed.) (1994). *Human error in medicine*. Hillsdale, NJ: Erlbaum.

6 Trade Indignation for Explanation

Indignation means you are upset or angry or annoyed at something that seems unreasonable. When you are faced with the rubble of a mishap, it is easy to get indignant. They should have looked out a little better! It is easy to get indignant about how other people misbehaved. If only they had followed the rules! But whenever you get indignant, you easily slip into counterfactual language. Whenever you get indignant, you easily regress into the Old View. Explaining human error falls by the wayside. Explaining human error becomes subordinate to getting upset about it. And people get upset about human error a lot.

I am looking at an Aviation Weekly, and its headline announces that "Computers continue to perplex pilots: Crash investigations again highlight prominence of human error and mode confusion".[1] *The writer of the Weekly is upset about the behavior of two pilots of a Boeing 737 who tried to get the autopilot engaged but did not succeed. The person who sent the Weekly to me was also upset. He was in fact indignant. "This is unacceptable!", he exclaimed. "How can we stop people from doing this?!"*

After taking off in the night, the pilots of the 737 had not noticed that the autopilot had not switched on. At the same time, the Captain got confused about which way the aircraft was turning. He rolled and rolled and rolled the wings of the aircraft all the wrong way. Perhaps he was mixing up his old Russian artificial horizon display with the American one he was looking at now. And the First Officer never lifted a finger to intervene. The aircraft crashed into the sea and everybody died.

Digging into the event, the Weekly found how it "tracked inexorably towards what would be a dumbfounding revelation". The Weekly observes how the First Officer "misunderstands" that you can't engage the autopilot if the wings aren't level and that this is dangerous if you're in a very black night over a featureless sea. But he tries anyway. The captain is flying and "evidently fails to register" what is going on, assuming "that the autopilot is engaged and has control". It does not, as the writer of the Weekly helpfully points out for us.

With "less than a minute before impact", the captain "misreads" the attitude indicator, "confusion is evident", and an "unhelpful reassurance" from the copilot only "serves to further mislead the captain". *"The pilot continues to roll the wrong way. The situation turns critical"* and then becomes *"irretrievable". The aircraft splashes into the sea.*

The Old View has its own logic, and it has its own language. The language of the old view dominates the example above. The writer puts himself outside the flow of events and looks back down on it. He sees how things are unfolding, "irretrievably" and "inexorably" towards a fatal *denouement.* He wonders how the people inside of the situation can be so confused and misled as to not see the outcome he already knows about. Despite the writer's growing conviction that this situation is going sour, the pilots he is watching trudge deeper and deeper into the confused, ill-coordinated, cognitive marsh where they will meet their deaths. He notes how the pilots inside the situation "fail to register" this, that, or the other thing, and how they send each other "unhelpful" signals on the way.

But all the writer can do is stand on the sideline and scratch his head at the spiraling, accelerating, escalating rush into lethal trouble that he has seen coming all along. The writer is counting down the seconds to impact, and then —
BOOM.
Told you so.

Indignation Trumps Explanation

Many stories about human error get written like the one above. You may have written or told stories about human error in the same way. But remember the tunnel picture from Chapter 3 and reproduced overleaf (See Figure 6.1). It illustrates where you stand in that case, just like the writer of the 737 story. You take the position of retrospective outsider (outside and hindsight), who sees the rattling advance toward an inevitable fatal outcome all the way. And you get upset about why people on the inside of the tunnel didn't see it the same way. Why they didn't see reality the way you now do.

The problem about taking the position of retrospective outsider is that it does not allow you to explain anything. From that position, all you can do is judge people for not noticing what you find so important now (given your knowledge of the outcome). From the position of retrospective outsider, it is possible only to condemn people for turning a manageable situation into an irretrievable one.

From this position, you will never be able to make sense of the behavior of those people. You will forever be relegated to asking rhetorical questions about

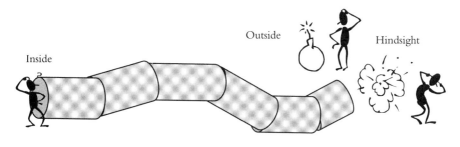

Figure 6.1 Different perspectives on a sequence of events: Looking from the outside and hindsight you have knowledge of the outcome and dangers involved. From the inside, you may have neither.

it. How could they not have seen? How could they fail to register? How could they mislead each other so much? From your perch up there, overlooking the sequence of events as it slithers its way to breakdown, all that will meet your questions is silence. Pure silence. The answers are not within your reach because you yourself have put yourself out of their range.

Remember that real understanding comes from putting yourself in the shoes of the people on the *inside* of the sequence of events; on the *inside* of the tunnel. Only there can you begin to make out why it made sense for people to do what they did. Because if it did, it will probably make sense for others too. And if it makes sense for others, then this type of breakdown may be more common than you think. Others could run into the same type of situation.

And all *you* did was get upset about it.

Take your pick: Be indignant or do something meaningful

It is of course understandable that you would get disturbed at the sight of apparently stupid behavior. Especially if it has tragic consequences. But the problem is that such indignation serves so little. The problem is that anger and indignation at a sequence of events that led to trouble is not the answer to making progress on safety.

There is a choice here. Do you want to feel angry and indignant about other people's performance? Then go right ahead, and talk in the language of the Old View. Adhere to its logic. Or do you want to do something meaningful in the wake of an event like this and help systems learn from failure and improve?

Take your pick. Either you keep on grumbling that people should not do ill-advised things like this, that they should be more competent, that this is unacceptable. Or you actually move forward. You shed the language, shake off the logic of the Old View, and do something to make progress on safety.

Remember the local rationality principle. What people do makes sense to them at the time, given their goals, knowledge and focus of attention. Who are you, then, to say that their behavior does not make sense? That it is "dumbfounding"? If their behavior does not make sense to you, that says something about you, about the perspective you have taken—not about them or their performance.

Old View and New View Language

Let us run through some of the key text of the 737 Old View example, and then, in contrast, through some of the key text of some other examples, including a New View one. As you will see:

- The first is the language of indignation, anger and incomprehension;
- The second is the language of understanding and explanation.

If you want to make progress on safety, which logic do you think you need to pursue? Which language should you use? Below are three examples, pried apart. They are the good, the bad and the ugly.

The bad

In the previous chapter, you saw how counterfactual language is the language of "should have", "could have", "if only". Those phrases are clearly out of order if you want to understand human error. Consider the following case, and then read on.

One June 10, 1995, a passenger ship named Royal Majesty *left St. Georges in Bermuda. On board were 1509 passengers and crewmembers who had Boston as their destination—677 miles away, of which more than 500 would be over open ocean. Innovations in technology have led to the use of advanced automated systems on modern maritime vessels. Shortly after departure, the ship's navigator set the ship's autopilot in the navigation (NAV) mode. In this mode, the autopilot automatically corrects for the effects of set and drift caused by the sea, wind and current in order to keep the vessel within a preset distance of its programmed track.*

Not long after departure, when the Royal Majesty *dropped off the St. Georges harbor pilot, the navigator compared the position data displayed by the GPS (satellite-based) and the Loran (ground/radio-based) positioning systems. He found that the two sets of data indicated positions within about a mile of each other—the expected accuracy in that part of the world. From there on, the* Royal Majesty *followed its programmed track (336 degrees), as indicated on the automatic radar plotting aid. The navigator plotted hourly fixes on charts of the area using position data from the GPS. Loran was used only as a back-up system, and when checked early on, it revealed positions about 1 mile southeast of the GPS position.*

About 34 hours after departure, the Royal Majesty *ran aground near Nantucket Island. It was about 17 miles off course. The investigation found that the cable leading from the GPS receiver to its antenna had come loose and that the GPS unit (the sole source of navigation input to the autopilot) had defaulted to dead-reckoning (DR) mode about half an hour after departure. Evidence about the loss of signal and default to DR mode was minimal, contained in a few short beeps and a small mode annunciation on a tiny LCD display meters from where the crew normally worked. In DR mode, there was no more correction for drift. A northeasterly wind had blown the Royal Majesty further and further west.*

In its report of the grounding of the *Royal Majesty*, the investigation board[2] says:

> Thus, had the officers regularly compared position information from the GPS and the Loran-C (another navigation device), they should not have missed the discrepant coordinates, particularly as the vessel progressed farther from its intended track.

"If only the officers had, then they should". This is the language of incomprehension and indignation. The language of judgments, of hindsight. It talks about a reality that never happened. It is, in other words, counterfactual. The phrase "particularly as the vessel progressed farther from its intended track" plays an interesting role here. It firmly places the report in the position of retrospective outsider. Only from *that* position is it visible that the divergence between actual and intended track is increasing. For the crew on the inside, there is no such divergence and no such trend at all. The vessel was not progressing farther from its intended track. It was ON track! Otherwise the crew would have done something differently.

If you want to understand human error, you have to put yourself on the inside of the tunnel and find out why it could have made sense for this crew to think that they were in the right place. In fact, you have to figure out why it made sense for them to not even ponder whether they were in the right place or not. The "cues" that could have shown the crew the real nature of the situation did not serve as cues to the crew in question at all. If you want to understand human error, your job is to find out why not. And do not rely on motivational explanations! You have to assume that people did not come to work to do a bad job (otherwise, don't turn to human factors as a field of explanation).

This crew, as most others, was motivated to not ground their vessel. Had they known about the outcome, they would have acted differently. The interesting question is how the crew was able to reconcile the various inputs they were getting with their developing understanding of the situation. Not how those cues, if only interpreted with as much hindsight knowledge as you now have, would have proved you right and them wrong.

Unfortunately, the report does not contain a good explanation for why reality developed the way it did (other than hints that the officers did not try hard enough to be good seamen). It would be interesting to probe deeper into how watchkeeping practices have evolved as a result of new technology on the ship bridge and why the crew's reliance on GPS made sense given their history with the technology. But instead there is more retrospective, counterfactual language. For example, upon sighting things that were not on the intended route, the report says how

> The second officer's response to these sightings should have been deliberate and straightforward. He should have been concerned as soon as the BB buoy was not sighted and then again when the lookouts sighted the red lights. Additionally, the second officer should have checked the Loran-C ….

The passage above is not only counterfactual (he should have this, he should have that). Notice how the report implicitly, insidiously puts up a bar here, a standard of what would be considered good seamanship, or professional behavior. Good seamen respond deliberately and straightforwardly to sightings. Good seamen get concerned as soon as there are contradictions or anomalies. Good seamen double-check with other sources.

Well, suppose that the report is right in its insinuations. Suppose that good seamen do all of that. Then it still has not proved that the sequence of events onboard this ship is evidence of *bad* seamanship. Perhaps the second officer would do all of that (and there is no reason to assume he wouldn't: he does not want to help ground a ship either) if he were in a doubtful situation that called for it. The interesting question, both psychologically and in terms of making progress on safety, is why the contradictions and anomalies that are now plain for all to see, were not interpreted as such at the time. Why the situation was not construed as sufficiently doubtful back then to warrant all that "good seamanship" behavior. How are normal, professional seamen liable to interpret cues (especially slightly ambiguous cues) in light of a worldview they believe to be true? And how do new navigation technologies interact with that age-old phenomenon?

This is what an investigation should begin to figure out. This is where an investigation can make real progress on safety. Because if this seaman did it, others might too. What is the larger phenomenon that places people like him at risk? What psychological trickery are they exposed to without realizing it? The report is silent on this. All it says, indignantly, is what the second officer (had he been a good seaman) should have done. As a result, there is little to no learning from failure that a community of practice takes away from a report like this one.

In summary, "bad" language does this:

- It says what people should or could have done rather than offering an explanation for why they did what they did;
- It implicitly puts up a standard of what is "good" practice and then judges the observed performance deficient by that standard.

The ugly

There are other ways in which you can get caught in indignation, rather than explanation. Let us go through some of the sections of the write-up about the 737 accident below. Very much in the beginning, the report says that:

> The three-man crew of the 737–300 completely lost the bubble—and everybody died. The illogicality of what happened there has left everybody in disbelief.

Almost all the words and phrases there breathe the spirit of the OldView. Begin with "completely lost the bubble". This is problematic at many levels. "Losing the bubble" is possible to hand down as a judgment only if you stand on the sidelines, knowing where a situation is really headed (what "the bubble" really is, or should be). There is no better way to reveal your position as a retrospective outsider than by using such language. "Losing the bubble" is the difference between what people inside the situation apparently knew and what they should have known (or what you now know). "Losing the bubble", moreover, is a folk model (see Chapter 13), an unspecific high-level label that everybody is supposed to understand (and consequently nobody really does). It is just saying "human error" in a more modern guise. It explains nothing.

Then, words like "illogicality" and "disbelief" really confirm the limitation of trying to understand performance from a position of retrospective outsider. If things look "illogical", they do so because you've got the wrong perspective! Trade places, once again, climb down from your retrospective perch and start looking inside the tunnel. If things look "illogical" they should not inspire "disbelief" and headscratching above and behind the sequence of events. They should inspire deeper investigation, a hunger for more insight, a thirst for making them look "logical"—as logical or rational as they must have looked to the people on the inside. For they must have: these pilots did not want to die. Even the writer of the 737 piece in the Aviation Weekly acknowledges that much:

> Suicidal actions by the pilot were ruled out and the process of scenario elimination tracked inexorably towards what would still be a dumb-founding revelation.

In "ugly" language about human error, words like "inexorable" appear. Words like "dumbfounding" appear. They are lexical labels for the tunnel cartoon. The inexorability attaches to the outside observer seeing the timebomb and wondering why people on the inside of the situation are not responding to it the way he would. The "dumbfounding" refers to the eventual explosion.

Figure 6.2   The language you use in your description of human error gives away where you stand. Here you clearly take the position of retrospective outsider.

In summary, when you use "ugly" language,

- You express disbelief and indignation at how people can fail to have seen the world the way you do now;
- You firmly put yourself in a position on the outside and in hindsight;
- You say, in so many words, "I told you so …". If only those people had seen the world the way you did, then the bad outcome wouldn't have had to happen.

Interestingly, even the 737 story pulls out to a wider perspective, and starts looking for a context in which the performance of the pilots could have made sense. Where everybody else may have done the same thing. Referring to the increasing bank angle (as the Captain kept rolling the wings of the aircraft the wrong way in his attempts to recover from the unintended turn), it says that

> The necessary dynamics of an unusual attitude recovery are likely to rapidly overextend even a mature airline pilot well beyond his experience and currency. You can talk about upset and loss of control all day. However, it is the mind-numbing physiological effects of the event itself …

In other words, other pilots are liable too. In fact, notice how the report implicitly puts up a standard of practice here. The dynamics required to recover from such a situation "are likely to rapidly overextend even a mature airline pilot well beyond his experience and currency". The report leaves unsaid what it means with "mature" and if it believes whether the pilot in question was thus "mature". But that no longer matters as much. Even if this pilot had been, the situation would likely still have taxed him beyond what he (and virtually any other pilot) could deliver.

The reasons the report offers include the "mindnumbing effects of the event itself", and then it proceeds to talk about different kinds of attitude instruments in aircraft that show, in different ways, how the aircraft is pitching and rolling relative to the horizon. It talks about fatigue and about instruments that are not co-located, robbing pilots of the ability to cross-check. It talks about a giant leap backward in autopilot switch design: from large paddle-like switches (from which it was easy to tell whether the autopilot had engaged or not), to press-buttons that do not reveal anything compelling on the newer 737s.

This language, the language of understanding and explanation, of context, is language that begins to open a door to progress on safety. What is it about spatial disorientation that makes pilots so vulnerable (even mature ones), especially if you combine it with the effects of fatigue, management and mental models of the autopilot, and transfer of experience across different instrument formats? Rather than trying to show, the whole time, how these three Bad Apples tracked inexorably towards their fatal outcome, the report actually turns around and probes how other pilots could be equally vulnerable to this same kind of scenario. That is good. That is potential progress on safety.

The good

So what counts as "good" language when it comes to talking about human error? Here is what it does:

- It avoids counterfactuals and judgmentals. It moves beyond obvious hindsight ("inexorable"), beyond indignation ("disbelief") and away from saying human error by any other name ("losing the bubble");
- But it needs to do more. It needs to offer an explanation for why it made sense for the people in question to do what they did. This is often the most difficult part. You need to rely on models and theory from for example psychology, sociology, organizational science, to build such an explanation. Chapters 14 and 15 will get you a bit on your way in this respect.

This is very much what the report into the Swissair 111 accident does.[3] As everybody else, investigators were interested in the actions of the pilots of the large passenger jet after the crew noticed smoke in the cockpit. A diversion airport (Halifax) was in their vicinity, but they did not make an emergency descent, and never made it there. Instead, the pilots took time sizing up the situation, going through checklists, and making preparations for fuel dumping to reduce their landing weight. The developing fire caught up with them and rendered the aircraft uncontrollable. It crashed into the sea, killing everybody onboard. That much is obvious in hindsight.

Now the interesting question is, what could the perspective of the crew have been? How could their actions have made sense? Let us look at the report.

> When the pilots started their descent toward Halifax at 0115:36, they had assessed that they were faced with an air conditioning smoke anomaly that did not require an emergency descent. Based on their perception of the limited cues available, they took steps to prepare the aircraft for an expedited descent, but not an emergency descent and landing.

Here the report deliberately tries to speak from the crew's perspective; it acknowledges that there were only limited cues available. Not only that, it makes sure that you understand that the crew took action on the basis of its "*perception of the limited cues available*". What mattered in understanding why the pilots did what they did, is *their* perception of the limited cues. Not what those cues would have meant to you, in hindsight, or what other cues might have been available that you now see but they didn't see back then. The report stays away from all of that. In its language, it does everything to keep looking inside the tunnel, from the position of the crew whose actions it is trying to understand.

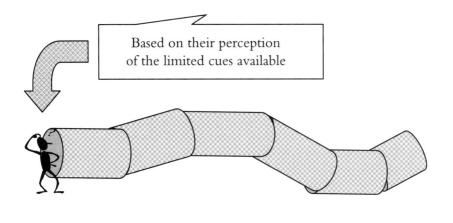

Figure 6.3 Using the right language, you can try to take the perspective of the people whose assessments and actions you are trying to understand.

On the basis of the pilots' understanding of the situation (they were faced with air conditioning smoke, a much less serious and slightly more common condition than fire), they did what any other crew would have done. In addition to this not being a fire in their worldview, why did the pilots not make a beeline for their diversion airport?

> The pilots were unfamiliar with the Halifax International Airport and did not have the approach charts readily available. The back-course instrument landing approach to Runway 06 was not pre-programmed into their flight management system. The pilots knew that they would have to take additional time to familiarize themselves with, and set up for, the approach and landing. They were given the weather information by the crew of an overflying aircraft, but did not know the runway lengths or orientation. Having runway and

> instrument approach information available is normal practice and is important in carrying out a safe approach and landing, particularly at an unfamiliar airport at night. ... The pilots also knew that the weight of the aircraft exceeded the maximum overweight landing limits for non-emergency conditions.

It makes sense. If you are unfamiliar with an airport (don't even know what runways there are, how long they are, whether they will carry or fit your airplane), you don't have the charts readily available, the approach is not programmed into the flight computer, it is a kind of approach that pilots do not do very often (if ever), and your aircraft is overweight, then it makes sense to think twice about where to put your priorities. Oh, and it is dark outside too, so visual references are not really an option.

As you see, the report not only tells us about the crew in question, and how they must have looked at the situation. It also puts this assessment in a larger context of canonical practice ("having information available is normal practice and important in carrying out a safe approach and landing"). Thus, the report converts a judgmental tactic of the Old View (micro-matching actual performance with supposed standards of good practice—remember Chapter 4?) into a powerful weapon of sensemaking. And then it supplies even more.

> In addition to these flight management circumstances, the pilots were aware that the meal service was underway, and that it would take some time to secure the cabin for a safe landing. Given the minimal threat from what they perceived to be air conditioning smoke, and the fact that there were no anomalies reported from the passenger cabin, they would likely have considered there to be a greater risk to the passengers and cabin crew if they were to conduct an emergency descent and landing without having prepared the cabin and positioned the aircraft for a stabilized approach and landing. It can be concluded that the pilots would have assessed the relative risks differently had they known that there was a fire in the aircraft.

There are more reasons not to start screaming down. First, there is no fire, only air-conditioning smoke. Then the airport, and the approach to it, are unfamiliar and unprepared for. Now the cabin is busy with a meal service, which makes a sudden, rapid descent risky. Also, the cabin does not report any anomalies. No reports of smoke or fire are coming from there.

In a final nod to the position of the retrospective outsider, the report acknowledges that this crew too, would have acted differently if they had known the seriousness of the situation. The point is, they did not. How could it have made sense for the crew to attain and confirm their interpretation, and to consequently do what they did? How would it make sense to other crews too? These are critical questions, and the Swissair 111 report goes a long way in addressing them meaningfully. This is good. This is where progress on safety begins to rise from the rubble of an accident.

Notes

1   Air Safety Week (2005, January 24). Vol. 19, No. 3. Washington, DC.
2   National Transportation Safety Board. (1997). *Grounding of the Panamanian passenger ship* Royal Majesty *on Rose and Crown shoal near Nantucket, Massachusetts, June 10, 1995.* (NTSB Rep. No. MAR-97/01). Washington, DC: Author.
3   Transportation Safety Board of Canada (2003). *Aviation investigation report: In-flight fire leading to collision with water. Swissair MD-11 HB-IWF, Peggy's Cove, Nova Scotia 5 nm SW, 2 September 1998* (Report Number A98H0003). Quebec: TSB.

7 Sharp or Blunt End?

Reactions to failure focus firstly and predominantly on those people who were closest to producing or potentially avoiding the mishap. It is easy to see these people as the engine of action. If it were not for them, the trouble would not have occurred.

Someone called me on the phone from London, wanting to know how it was possible that train drivers ran red lights. Britain had just suffered one of its worst rail disasters—this time at Ladbroke Grove near Paddington station in London. A commuter train had run head-on into a high-speed intercity coming from the other direction. Many travelers were killed in the crash and ensuing fire. The investigation returned a verdict of "human error". The driver of the commuter train had gone right underneath signal 109 just outside the station, and signal 109 had been red, or "unsafe". How could he have missed it? A photograph published around the same time showed sensationally how another driver was reading a newspaper while driving his train.

Blunt End and Sharp End

In order to understand error, you have to examine the larger system in which these people worked. You can divide an operational system into a sharp end and a blunt end:

- At the sharp end (for example the train cab, the cockpit, the surgical operating table), people are in direct contact with the safety-critical process;
- The blunt end is the organization or set of organizations that supports and drives and shapes activities at the sharp end (for example the airline or hospital; equipment vendors and regulators).

The blunt end gives the sharp end resources (for example equipment, training, colleagues) to accomplish what it needs to accomplish. But at the same time it puts on constraints and pressures ("don't be late, don't cost us any unnecessary money, keep the customers happy"). Thus the blunt end shapes, creates, and can even encourage opportunities for errors at the sharp end. Figure 7.1 shows this flow of influences through a system. From blunt to sharp end; from upstream to downstream; from distal to proximal. It also shows where we typically aim our reactions to failure: at the people at the sharp end.

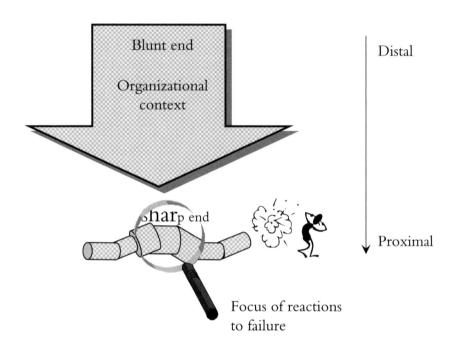

Figure 7.1 Failures can only be understood by looking at the whole system in which they took place. But in our reactions to failure, we often focus on the sharp end, where people were closest to causing or potentially preventing the mishap.

Why Do You Focus on the Sharp End?

Looking for sources of failure far away from people at the sharp end is counterintuitive. And it can be difficult. If you find that sources of failure lie really at the blunt end, this may call into question beliefs about the safety of the entire system. It challenges previous views. Perhaps things are not as well-organized or well-designed as people had hoped. Perhaps this could have happened any time. Or worse, perhaps it could happen again.

The Ladbroke Grove verdict of "driver error" lost credibility very soon after it came to light that signal 109 was actually a cause célèbre *among train drivers. Signal 109 and the entire cluttered rack on which it was suspended together with many other signals, were infamous. Many drivers had passed an unsafe signal 109 over the preceding years and the drivers' union had been complaining about its lack of visibility.*

In trains like the one that crashed at Ladbroke Grove, automatic train braking systems (ATB) had not been installed because they had been considered too expensive. Train operators had grudgingly agreed to install a "lite" version of ATB, which in some sense relied as much on driver vigilance as the red light itself did.

Reducing Surprise by Pinning Failure on Local Miscreants

Some people and organizations see surprise as an opportunity to learn. Failures offer them a window through which they can see the true internal workings of the system that produced the incident or accident. These people and organizations are willing to change their views, to modify their beliefs about the safety or robustness of their system on the basis of what the system has just gone through. This is where real learning about failure occurs, and where it can create lasting changes for the good.

But such learning does not come easy. And it does not come often. Challenges to existing views are generally uncomfortable. Indeed, for most people and organizations, coming face to face with a mismatch between what they believed and what they have just experienced is difficult. These people and organizations will do anything to reduce the nature of the surprise.

Some fighter pilots are not always kind on the reputation of a comrade who has just been killed in an accident. Sociologists have observed how his or her fellow pilots go to the bar and drink to the fallen comrade's misfortune, or more likely his or her screw-up, and put the drinks on his or her bar tab. This practice is aimed at highlighting or inventing evidence for why he or she wasn't such a good pilot after all. The transformation from "one of us" into "bad pilot" would psychologically shield those who do the same work from realizing their equal vulnerability to failure.

People and organizations often want the surprise in the failure to go away, and with it the challenge to their views and beliefs. The easiest way to do this is to see the failure as something local, as something that is merely the problem of a few individuals who behaved in uncharacteristic, erratic or unrepresentative (indeed, locally "surprising") ways.

Potential revelations about systemic vulnerabilities were deflected by pinning failure on one individual in the case of November Oscar. 1 November Oscar was an older Boeing 747 "Jumbojet". It had suffered earlier trouble with its autopilot, but on this morning everything else conspired against the pilots too. There had been more headwind than forecast, the weather at the destination was very bad, demanding an approach for which the co-pilot was not qualified but granted a waiver, while he and the flight engineer were actually afflicted by gastrointestinal infection. Air traffic control turned the big aircraft onto a tight final approach, which never gave the old autopilot enough time to settle down on the right path. The aircraft narrowly missed a building near the airport, which was shrouded in thick fog. On the next approach it landed without incident.

November Oscar's captain was taken to court to stand trial on criminal charges of "endangering his passengers" (something pilots do every time they fly, one fellow pilot quipped). The case centered around the crew's "bad" decisions. Why hadn't they diverted to pick up more fuel? Why hadn't they thrown away that approach earlier? Why hadn't they gone to another arrival airport? These questions trivialized or hid the organizational and operational dilemmas that confront crews all the time. The focus on customer service and image; the waiving of qualifications for approaches; putting more work on qualified crewmembers; heavy traffic around the arrival airport and subsequent tight turns; trade-offs between diversions in other countries or continuing with enough but just enough fuel. And so forth.

The vilified captain was demoted to co-pilot status and ordered to pay a fine. He later committed suicide. The airline, however, had saved its public image by focusing on a single individual who—the court showed—had behaved erratically and unreliably.

Potentially disruptive lessons about the system as a whole are transformed into isolated hiccups by a few uncharacteristically ill-performing individuals. This transformation relieves the larger organization of any need to change views and beliefs, or associated policies or spending priorities. The system is safe, if only it weren't for a few unreliable humans in it.

Faced with a bad, surprising event, we change the event or the players in it, rather than our beliefs about the system that made the event possible. Instead of modifying our views in the light of the event, we re-shape, re-tell and re-inscribe the event until it fits the traditional and non-threatening view of the system. As far as organizational learning is concerned, the mishap might as well not have happened. The proximal nature of our reactions to failure makes that expensive organizational lessons can go completely unlearned.

The pilots of a large military helicopter that crashed on a hillside in Scotland in 1994 were found guilty of gross negligence. The pilots did not survive—29 people died in total—so their side of the story could never be heard. The official inquiry had no problems with "destroying the reputation of two good men", as a fellow pilot put it. Indeed, many other pilots felt uneasy about the conclusion. Potentially fundamental vulnerabilities (such as 160 reported cases of Uncommanded Flying Control Movement or UFCM in computerized helicopters alone since 1994) were not looked into seriously.[2]

Really understanding system safety and risk not only begins with calling off the hunt for Bad Apples or their errors. It necessarily coincides with an embrace of the system view—seeing both blunt end and sharp end and how they interact to shape practice and define what is normal or expected throughout an organization. This means zooming out, away from looking just at the sharp end, and incorporating blunt end policies and priorities and how these help determine people's goals and practices at the sharp end.

Notes

1   Wilkinson, S. (1994). The November Oscar Incident. *Air and Space*, February–March.
2   *Sunday Times*, June 25, 2000.

8 You Can't Count Errors

Many people believe that getting a grip on their "human error problem" means quantifying it. If only they could put a number on it, then they would have a better idea of how much people violate, how often they go wrong, and where. Managers typically like these numbers. In fact, management often occurs by numbers. So why not safety? We all think we know that a stubborn 70 per cent of mishaps is still due to human error. So if you can make the error count go down, your system has becomes safer. Right?

Wrong. As Richard Cook reminds us, it is a mere myth that 70 per cent represents the distance we have to go before we reach full safety. This myth suggests that full safety lies somewhere on, or beyond, the horizon, and that 70 per cent human errors are between you and that goal. But this assumption about the location of safety is an illusion, and efforts to measure the distance to it are like measuring your distance from a mirage (Figure 8.1).

Safety is right here, right now, right under your feet—not yonder across some 70 per cent. You don't have to traverse 70 per cent first to get to 0 per cent error and 100 per cent safety. This really is one of the basic tenets of the Old View. It assumes that safety, once established, can be maintained by keeping human performance within established boundaries. So the point would be to find ways to limit human variability in otherwise well-designed and safe systems.

But this is not how it works. People in complex systems *create* safety. They make it their job to anticipate forms of, and pathways toward, failure. They invest in their own resilience and that of their system by tailoring their tasks, by inserting buffers, routines, heuristics, double-checks, memory aids. The occasional human contribution to failure occurs because complex systems need an overwhelming human contribution for their safety. Human error is the inevitable by-product of the pursuit of success in an imperfect, unstable, resource-constrained world.

Error counting supposedly tells you something essential about the safety of your system. Today, you have a choice of many error counting and categorization systems. Some are tailored to specific settings (such as air traffic control), others rely more on general models of psychology or organizational failure. What they have in common is numerology, the belief that by reading the runes, by divining

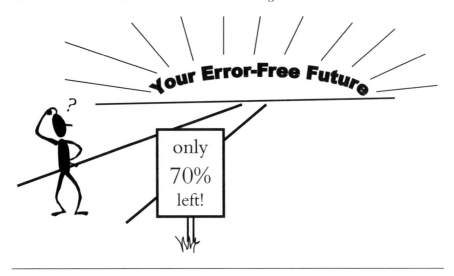

Figure 8.1 The mirage on the horizon. We think all we need to do to reach perfect safety is get rid of the last 70 per cent human errors. But opportunities for safety are right here, right under our feet. Original metaphor by Richard Cook.

the stars, you can colonize and control a part of your future. As it did in ancient times, this probably helps defray some of our deepest human anxieties. But it offers nothing in the way of real progress on safety. Here is why.

It is hard to agree on what an "error" is

Tabulation of errors may have worked once upon a time, when scientists set up tightly controlled laboratory studies to investigate human performance. In these lab studies, they shrunk human tasks and opportunities to err down to a bare minimum. Then they could count single, measurable errors as a basic unit of human performance. This kind of experimentation left the scientist with spartan but quantifiable results.

When it comes to human error "in the wild"—as it occurs in natural complex settings—such tabulation is impossible, as "errors" themselves become almost indistinguishable from the noisy, busy background of real work. What people see as "erroneous" depends very much on the perspective they themselves have. It is almost impossible for people to reach agreement on what counts as an "error".

Consider a study reported by Erik Hollnagel and René Amalberti,[1] whose purpose was to test an error measurement method for air traffic controller "errors". The method asked observers to count errors and categorize the types of errors using a taxonomy proposed by the developers. It was tested by pairs of psychologists and air traffic controllers who studied air traffic control (ATC) work going on in real time. The observing air traffic controllers and psychologists, both trained in the error taxonomy, were instructed to take note of all the errors they could see.

Despite common indoctrination, there were substantial differences between the numbers and kinds of errors each of the two groups of observers noted, and only a very small number of errors were actually observed by both. People watching the same performance, using the same tool to classify behavior, came up with totally different error counts! Closer inspection of the score sheets revealed that air traffic controllers and psychologists tended to use different subsets of the error types available in the tool, indicating just how negotiable the notion of error is. Air traffic controllers relied on external working conditions (e.g. interfaces, personnel and time resources) to refer to, and categorize errors, whereas psychologists preferred to locate the error somewhere in presumed quarters of the mind (e.g. working memory), or in some mental state (e.g. attentional lapses).

Moreover, air traffic controllers who actually did the work could tell the error coders that they both had it wrong. Debriefing sessions exposed how many observed "errors" were not errors at all to those "committing" them, but rather normal work; part of deliberate strategies intended to manage problems or foreseen situations that the error counters had neither seen, nor understood as such if they had.

This is typical for error counting instruments: they claim to be able pick up more "errors" than the observed practitioners themselves. Some may argue that this makes it a strong method, a success for scientific objectivity and management-by-numbers. But it is more likely hubris.

In the case above, for example, early transfers of aircraft were not an error, but turned out to correspond to a deliberate strategy connected to a controller's foresight, planning ahead and workload management. Rather than an expression of a weakness, such strategies uncovered sources of robustness that would never have come out, or would even have been misrepresented and mischaracterized, with just the data in the counting method. The error counters were forced to wipe most of their observed "errors" from their slates, leaving them empty of any meaningful data.

Now suppose that these "errors" would have been left in the results of the counting exercise. The observed air traffic controllers would have come out looking pretty badly, as they made so-and-so many errors of perception, and so-

and-so many errors of prediction or memory or situation assessment. Managers may not have been pleased.

This is a big risk of counting and categorizing errors, certainly without checking with the observed people themselves. Remember the local rationality principle: what people do makes sense to them given the circumstances at the time—given their goals, their attention and knowledge, their understanding of the situation. In other words, knowledge of context is critical to understanding error. Answers to why people do what they do often lie in the context surrounding their actions (an early transfer of an airplane is not an error: it is a sensible strategy of workload management given the unfolding context). Counting errors and stuffing them away in a measurement instrument (and handing the results to management in the form of a bar chart) removes that context. It is gone, no longer there.

- Without context, you cannot reconstruct local rationality.
- Without local rationality, you cannot understand human error.
- So counting error is contrary to understanding error.

Human Error—In The Head or The World?

One thing that error categorization tools often do is lead observers through various steps to help them explain the error. What may have caused it, where did the error come from? This is a good intention. Errors, after all, are not explanations of failure, but demand explanations.

There seems to be a basic choice that observers have when they pick their categorization tool. It allows them to trace the reasons for the error back to:

- The head (of the person committing the error);
- The world (in which the person committing the error found him- or herself).

But there are pitfalls in both.

The categorization tool for the analysis of human errors in air traffic control mentioned above takes the analyst through a series of questions that are based on an elaborate information processing model of the human mind. It begins with perceptual processes and points the analyst to possible problems or difficulties there. Then it goes on along the processing pathway, hoping to guide the analyst to the source of trouble in a long range of psychological processes or structures: short-term memory, long-term memory, decision making, response selection, response execution, and even the controller's image of him or herself. For each observed error, the journey through the questions can be long and arduous and the final destination (the supposed source of error) dubious and hard to verify.

These human error analyses deal with the complexity of behavior by simplifying it down to boxes; by nailing the error down to a psychological process or structure. For example, it was an error of perceptual store, or one of working memory, or one of judgment or decision making, or one of response selection. The aim is to conclude that the error originated in a certain stage along a psychological processing pathway in the head. These approaches basically explain error by taking it back to the mind from which it came.

The shortcomings, as far as investigating human error is concerned, are severe. These approaches hide an error back in the mind under a label that is not much more enlightening than "human error" is. In addition, the labels made popular in these approaches (such as working memory or response execution) are artifacts of the language of a particular psychological model. This model may not even be right, but it sure is hard to prove wrong. Who can prove the existence of short-term memory? But who can prove that it does not exist?

Explaining human error on the basis of internal mental structures will leave other people guessing as to whether the investigator was right or not. Nobody can actually see things like short-term memories or perceptual stores, and nobody can go back into the short-term memories or perceptual stores of the people involved to check your work. Other people can only hope you were right when you picked the psychological category.

By just relabeling human error in more detailed psychological terms, you remain locked in a practice where anyone can make seemingly justifiable, yet unverifiable assertions. Such supposed explanations of error remain fuzzy and uncertain and inconclusive, and low on credibility.

Some error categorization tools see this shortcoming, and try to locate the reason for the error in the world surrounding the person who committed the error. They will take the analyst by the hand and lead him or her through the warren of organizational reality.

One error categorization tool, for example, explains that operators make errors because their supervision was deficient, or their equipment design inadequate. But this simply replaces one error with another. Instead of "operator error", we now say "supervisor error" or "designer error".

Explaining one error by pointing to another is really not an explanation at all. It does nothing to get away from the Old View. It doesn't really help locate part of the explanation for error in the situation that surrounded people. It simply puts the error in somebody else's hands, or in somebody else's head. Another Bad Apple.

So should you look for the source of error in the head or in the world? The first alternative is used in various human error coding tools, and in fact often implied in investigations. For example, when you use "complacency" as a label to explain behavior, you really look for how the problem started with an individual who was not sufficiently motivated to look closely at critical details of his or her situation.

As said before, such an approach to "explaining" human error is a dead-end. It prevents you from finding enduring features of the operational environment that actually produce the controversial behavior (and that will keep producing it if left in place). The assumption that errors start in the head also leaves your conclusion hard to verify for others.

The alternative—look for the source of error in the world—is a more hopeful path. Human error is systematically linked to features of the world—the tasks and tools that people work with, and the operational and organizational environment in which people carry out that work. If you start with the situation (and resist simply finding other "errors" in that situation), you can identify, probe and document the reasons for the observed behavior, without any need to resort to non-observable processes or structures or big labels in someone's head.

To "reverse engineer" human error, later chapters will encourage you to reconstruct how people's mindset unfolded and changed over time. You would think that reconstructing someone's unfolding mindset begins with the mind. The mind, after all, is the obvious place to look for the mindset that developed inside of it. Was there a problem holding things in working memory? What was in the person's perceptual store? Was there trouble retrieving a piece of knowledge from long-term memory? These are indeed the kinds of questions

To understand what went on in someone's mind, you have to reconstruct the situation in which the mind found itself

asked in some human error analysis tools and incident reporting systems. But to begin to understand what may have gone on in somebody's mind, you'd do better by first reconstructing the situation in which the mind found itself.

Human Error—Matter over Mind

When you want to understand human error, it makes sense to start with the situation:

- Past situations can be reconstructed to a great extent, and documented in detail. They can be traced by other people in a way that assertions about unobservable psychological mechanisms cannot;
- There are tight and systematic connections between situations and behavior; between what people did and what happened in the world around them.

These connections between situations and behavior work both ways:

- People change the situation by doing what they do; by managing their processes;
- But the evolving situation also changes people's behavior. An evolving situation provides changing and new evidence; it updates people's understanding; it presents more difficulties; it forecloses or opens pathways to recovery.

You can uncover the connections between situation and behavior, investigate them, document them, describe them, represent them graphically. Other people can look at the reconstructed situation and how you related it to the behavior that took place inside of it. Other people can actually trace your explanations and conclusions. Starting with the situation brings efforts to understand human error out in the open. It does not rely on hidden psychological structures or processes, but instead allows verification and debate by those who understand the domain.

A large part of understanding human error, then, is about understanding in detail the situation in which the human was working; about the tasks he or she was carrying out; about the tools that were used. The reconstruction of mindset begins not with the mind. It begins with the circumstances in which the mind found itself. From those, you can show:

- How the process, the situation, changed over time;
- How people's assessments and actions evolved in parallel with their changing situation;

- How features of people's tools and tasks and their organizational and operational environment influenced their assessments and actions inside that situation.

Let Go of the Mind-Matter Divide Altogether

So does the source of human error lie in the world or in the head? The real answer lies somewhere in the middle. In fact, if you really want to understand human performance, you will have to let go of the sharp distinction between somebody's head, or mind, and the situation in which that mind found itself.

In an unfolding situation, what people are thinking, concluding, planning, deriving, assessing, is inextricably bound up with how circumstances are progressing. This in turn influences where people will look, what goals they likely pursue and where they may help steer circumstances next. The relationship is reciprocal and ever-ongoing, and understanding somebody's actions or assessments just by looking into the head is impossible.

Letting go of the divide between mind and world, and tracing how people's understanding of a situation developed hand in hand with unfolding circumstances, is key to a NewView analysis of error. This becomes very obvious when you start looking for "causes", as in the next chapter. The first thing to do is to stop asking the question whether it was human error or mechanical failure.

Notes

1 Hollnagel, E. and Amalberti, R. (2001). The emperor's new clothes: Or whatever happened to "human error"? In S.W.A. Dekker (ed.), *Proceedings of the 4th International Workshop on Human Error, Safety and Systems Development*. Linköping Sweden: Linköping University, pp. 1–18.

9 Cause is Something You Construct

What is the cause of the mishap? In the aftermath of failure, no question seems more acute. There can be significant pressure from all kinds of directions to pinpoint a cause:

- Not knowing what made a system fail is really scary;
- People want to start investing in countermeasures;
- People want to know how to adjust their behavior to avoid the same kind of trouble;
- People may simply seek retribution, punishment, justice.

Two persistent myths drive our search for the cause of failure:

- We think we can make a distinction between human error and mechanical failure, and that a mishap has to be caused by either one or the other. This is an oversimplification. Once you acknowledge the complexity of failure, you will find that the distinction between mechanical and human contributions becomes very blurred; even impossible to maintain. The first part of this chapter talks about that.

- We think there is something like *the* cause of a mishap (sometimes we call it the root cause, or primary cause), and if we look in the rubble hard enough, we will find it there. The reality is that there is no such thing as *the* cause, or primary cause or root cause. Cause is something we construct, not find. And how we construct causes depends on the accident model that we believe in. The second part of this chapter talks about that.

Human Error or Mechanical Failure?

Was this mishap due to human error, or did something else in the system break? You hear the question over and over again—in fact, it is often the first question people ask. The question, however, demonstrates an oversimplified belief in the roots of failure. And it only very thinly disguises the bad apple theory: the system is basically safe, but it contains unreliable components. These components are either human or mechanical, and if one of them fails, a mishap ensues.

Early investigation can typically show that a system behaved as designed or programmed, and that there was nothing mechanically wrong with it. This is taken as automatic evidence that the mishap must have been caused by human error—after all, nothing was wrong with the system. If this is the conclusion, then the Old View of human error has prevailed. Human error causes failure in otherwise safe, well-functioning systems. The reality, however, is that the "human error" does not come out of the blue. Error has its roots in the system surrounding it; connecting systematically to mechanical, programmed, paper-based, procedural, organizational and other aspects to such an extent that the contributions from system and human begin to blur.

Passenger aircraft have "spoilers"—panels that come up from the wing on landing, to help brake the aircraft during its roll-out. Before landing, pilots have to manually "arm" them by pulling a lever in the cockpit. Many aircraft have landed without the spoilers being armed, some cases even resulting in runway overruns. Each of these events gets classified as "human error"—after all, the human pilots forgot something in a system that is functioning perfectly otherwise.

But deeper probing reveals a system that is not at all functioning perfectly. Spoilers typically have to be armed after the landing gear has come out and is safely locked into place. The reason is that landing gears have compression switches that tell the aircraft when it is on the ground. When the gear compresses, it means the aircraft has landed. And then the spoilers come out (if they are armed, that is). Gear compression, however, can also occur while the gear is coming out in flight, because of air pressure from the slipstream around a flying aircraft, especially if landing gears fold open into the wind. This could create a case where the aircraft thinks it is on the ground, when it is not. If the spoilers would already be armed at that time, they would come out too—not good while still airborne. To prevent this, aircraft carry procedures for the spoilers to be armed only when the gear is fully down and locked. It is safe to do so, because the gear is then orthogonal to the slipstream, with no more risk of compression.

But the older an aircraft gets, the longer a gear takes to come out and lock into place. The hydraulic system no longer works as well, for example. In some aircraft, it can take up to half a minute. By that time, the gear extension has begun to intrude into other cockpit tasks that need to happen—selecting wing flaps for landing; capturing and tracking the electronic glide slope towards the runway; and so forth. These are items that come after the "arm spoilers" item on a typical before-landing checklist. If the gear is still extending, while the world has already pushed the flight further down the checklist, not arming the spoilers is a slip that is easy to make.

Combine this with a system that, in many aircraft, never warns pilots that their spoilers are not armed; a spoiler handle that sits over to one, dark side of the center cockpit console, obscured for one pilot by power levers, and whose difference between armed and not-armed may be all of one inch, and the question becomes: is this mechanical failure or human error?

One pilot told me how he, after years of experience on a particular aircraft type, figured out that he could safely arm the spoilers four seconds after "gear down" was selected, since the critical time for potential gear compression was over by then. He had refined a practice whereby his hand would go from the gear lever to the spoiler handle slowly enough to cover four seconds—but it would always travel there first. He thus bought enough time to devote to subsequent tasks such as selecting landing flaps and capturing the glide slope. This is how practitioners create safety: they invest in their understanding of how systems can break down, and then devise strategies that help forestall failure.

The deeper you dig, the more you will understand why people did what they did, based on the tools and tasks and environment that surrounded them. The further you push on into the territory where their errors came from, the more you will discover that the distinction between human and system failure does not hold up.

The Construction of Cause

Look at two official investigations into the same accident. One was conducted by the airline whose aircraft crashed somewhere in the mountains. The other was conducted by the civil aviation authority of the country in which the accident occurred, and who employed the air traffic controller in whose airspace it took place.

The authority says that the controller did not contribute to the cause of the accident, yet the airline claims that air traffic control clearances were not in accordance with applicable standards and that the controller's inadequate language skills and inattention were causal. The authority counters that the pilots' inadequate use of flightdeck automation was actually to blame, whereupon the airline points to an inadequate navigational database supplied to their flight computers among the causes. The authority explains that the accident was due to a lack of situation awareness regarding terrain and navigation aids, whereas the airline blames lack of radar coverage over the area. The authority states that the crew failed to revert to basic navigation when flight deck automation usage created confusion and workload, whereupon the airline argues that manufacturers and vendors of flightdeck automation exuded overconfidence in the capabilities of their technologies and passed this on to pilots. The authority finally blames ongoing efforts by the flight crew to expedite their approach to the airport in order to avoid delays, whereupon the airline lays it on the controller for suddenly inundating the flight crew with a novel arrival route and different runway for landing.[1]

Table 9.1 Two statements of cause about the same accident

Causes according to Authority:	Causes according to Airline:
Air Traffic Controller did not play a role	No standard phraseology, inadequate language and inattention by Controller
Pilots' inadequate use of automation	Inadequate automation database
Loss of pilots' situation awareness Failure to revert to basic navigation	Lack of radar coverage over area Overconfidence in automation sponsored by vendors
Efforts to hasten arrival	Workload increase because of Controller's sudden request

So who is right? The reality behind the controversy, of course, is that both investigations are right. They are both right in that all of the factors mentioned were in some sense causal, or contributory, or at least necessary. Make any one of these factors go away and the sequence of events would probably have turned out differently. But this also means that both sets of claims are wrong. They are both wrong in that they focus on only a subset of contributory factors and pick and choose which ones are causal and which ones are not. This choosing can be driven more by socio-political and organizational pressures than by the evidence found in the rubble.

Cause is not something you find. Cause is something you construct. How you construct it and from what evidence, depends on where you look, what you look for, who you talk to, what you have seen before, and likely on who you work for.

There is No "Root" Cause. It is Up to You

How is it that a mishap gives you so many causes to choose from? Part of the story is the sheer complexity of the systems that we have put together. But part of it is also that we have protected our systems so well against failure. A lot needs to go wrong for an incident or accident to occur. The potential for risk, in many industries, has been recognized long ago, and efforts to manage that risk are ongoing. Consequently, major investments have been made in protecting them against the breakdowns that we know or think can occur. These so-called "defenses" against failure contain human and engineered and organizational elements.

Flying the right approach speeds for landing while an aircraft goes through its subsequent configurations (of flaps and slats and wheels that come out), is safety-critical. As a result it has evolved into a well-defended process of double-checking and cross-referencing between crew members, speed booklets, aircraft weight, instrument settings, reminders and call-outs, and in some aircraft even by engineered interlocks.

Accidents can occur only if multiple factors succeed in eroding or bypassing all these defenses. The breach of any of these layers can be "causal". For example, the crew opened the speed booklet on the wrong page (i.e. wrong aircraft landing weight). But this fails to explain the entire breakdown, because other layers of defense had to be broken or side-stepped too. Why did the crew open the booklet on the wrong page? What is the cause of that action? Was it their expectation of aircraft weight based on fuel used on that trip; was it a misreading of an instrument? And once pinpointed, what is the cause of that cause? And so forth.

You can find "causes" of failures everywhere. The causal web quickly multiplies and fans out, like cracks in a window. What you call "root cause" is simply the place where you stop looking any further. You see that factor as necessary for the mishap to happen. Not only that, you also deem that "cause" sufficient. Nothing else would have needed to go wrong, otherwise you would also have to label those things as "causes".

Of course, many industries find ways around that by offering investigators the possibility of labeling causes and contributory factors. Or they let you hedge by identifying "probable" causes. By saying "probable", you indicate that you are not sure, that your conclusions are tentative. The aviation industry typically deals in "probable cause", but this has become such a well-worn phrase that it has all but lost its original meaning. Even the "probable causes" of aircraft accidents are of necessity:

- Selective. There are only so many things you can label "causal" before the word "causal" becomes meaningless;
- Exclusive. They leave out factors that were also necessary and only jointly sufficient to "cause" the failure;
- Oversimplifications. They highlight only a few hotspots in a long, twisted and highly interconnected, much finer-grained web that spreads far outside the causes you come up with.

As far as the rubble is concerned, there are no such things as root or primary or probable causes or contributory factors. In fact, there is no end or origin anywhere (other than the Big Bang).

If you find a root or primary or probable cause or contributory factor, it was your decision to distinguish some things in the dense causal mesh by those labels. So it may actually be better to think in terms of *explanations* than causes. The real explanation is that all these factors came together.

But What if you *Have* to Identify "Probable Causes"?

If protocol prescribes that you must identify probable causes, one way to deal with the problems above is to generate, as "probable cause", a brief narrative; the shortest possible summary of issues and events surrounding a mishap that, in your understanding, were really important. Although also an oversimplification, this mitigates selectivity and exclusion by highlighting all kinds of factors without assigning some artificial after-the-fact ranking. But there are other strategies too.

The report into the in-flight fire and subsequent crash of a Swissair MD-11 off Halifax, Nova Scotia, humbly calls its conclusions "Findings as to causes and contributory factors". It has no fewer than seven pages of careful meanderings through the complex, interleaving stories of electrical wiring, flammability, chimney effects, insulation, arcing failure modes, and crew actions in reponse to ambiguous evidence. It also produces "Findings as to risk", showing how factors that lie deeply buried inside the aviation system can still present a hazard, even if supposed "causes" are dealt with. This includes, for example, the finding that the industry does not know much about how contamination affects an aircraft's continuing airworthiness.²

If protocol in aviation would not demand investigators to identify "probable causes" (which it now does, in its so-called Annex 13 to the ICAO convention), then the industry could end up with better explanations of accidents and perhaps even more meaningful ways forward to manage risk.

Some investigators get so frustrated with the need to present a "probable cause" that they simply pick the last technical event that, if removed, could have prevented the mishap from occurring. They end up with ridiculously short "causes" that are almost impossible to refute because *that* particular factor was certainly necessary to have the outcome occur. Sufficiency is no longer a criterion for including a cause or contributory factor, as it would generate a causal list at least as long as the investigation report itself.

Aircraft sometimes settle on their tail when they are loaded incorrectly. Loading an aircraft can be tricky business, as load and balance sheets are increasingly computerized, based on certain assumptions (e.g. how much a mature passenger weighs); produced by people who will neither see the load, nor put it on onboard, nor fly the aircraft and who may sit a city or even continent away from where the aircraft is parked.

The organizational story behind loading is so complex that one investigator (after trying to tell that story) simply wrote "the incident was caused by the aircraft's center of gravity being outside its aft limit". And that was it. Nobody could really disagree with this causal conclusion either.

But rather than saying, "this is what I believe is the cause", the investigator is perhaps saying "I believe I am constrained by having to identify probable causes! This is a very complex story and I have dug into it for months or years." It is a sign of protest. It could also be a sign of resignation.

Indeed, acknowledging the traps inherent in the construction of cause, some people avoid labeling causes altogether. And they do come up with better explanations.

In a break with the tradition of identifying "probable causes" in aviation crashes—which oversimplify the large and intertwined web surrounding a failure—Judge Moshansky's investigation of the Air Ontario crash at Dryden, Canada in 1989 did not produce any probable causes. The pilot in question had made a decision to take off with ice and snow on the wings, but, as Moshanky's commission wrote, "that decision was not made in isolation. It was made in the context of an integrated air transportation system that, if it had been functioning properly, should have prevented the decision to take off... there were significant failures, most of them beyond the captain's control, that had an operational impact on the events at Dryden ... regulatory, organizational, physical and crew components ...."

Instead of forcing this complexity into a number of probable causes, the Commission generated 191 recommendations which pointed to the many "causes" or systemic issues underlying the accident on that day in March 1989. Recommendations ranged in topic from the introduction of a new aircraft type to a fleet, to management selection and turnover in the airline, to corporate take-overs and mergers in the aviation industry.[3]

There is No Single Cause

So what is the cause of the accident? This question is just as bizarre as asking what the cause is of not having an accident. There is no single cause. Neither for failure, nor for success. In order to push a well-defended system over the edge (or make it work safely), a large number of contributory factors are necessary and only jointly sufficient.

So where you focus in your search for cause is something that the evidence in a mishap will not necessarily determine for you. It is up to your investigation. Erik Hollnagel lucidly defines cause as "the identification, after the fact, of a limited set of aspects of the situation that are seen as necessary and sufficient conditions for the observed effect(s) to have occurred. The cause, in other words, is constructed rather than found."[4]

What is interesting, then, is that the identification of "cause" says more about you than about the mishap. And one of the things it says about you is what accident model you believe in.

Notes

1. See: Aeronautica Civil (1996). *Aircraft Accident Report: Controlled flight into terrain American Airlines flight 965, Boeing 757–223, N851AA near Cali, Colombia, December 20, 1995.* Santafe de Bogota, Colombia: Aeronautica Civil Unidad Administrativa Especial, and American Airlines' (1996) Submission to the Cali Accident Investigation.
2. Transportation Safety Board of Canada (2003). *Aviation investigation report: In-flight fire leading to collision with water. Swissair MD–11 HB-IWF, Peggy's Cove, Nova Scotia 5 nm SW, 2 September 1998* (Report Number A98H0003). Quebec: TSB.
3. Moshansky, V.P. (1992). *Commission of inquiry into the Air Ontario accident at Dryden, Ontario* (Final report, vol. 1–4). Ottawa, ON: Minister of Supply and Services, Canada.
4. Hollnagel, E. (2004). *Barriers and accident prevention.* Aldershot, UK: Ashgate Publishing Co, p. 33.

10 What is Your Accident Model?

Where you look for causes depends on how you believe accidents happen. Whether you know it or not, you apply an *accident model* to your analysis and understanding of failure. An accident model is a mutually agreed, and often unspoken, understanding of how accidents occur. Today there are, roughly, three kinds of accident models:[1,2]

- **The sequence-of-events model.** This model sees accidents as a chain of events that leads up to a failure. One event causes another, and so on, until the entire series produces an accident. It is also called the domino model, as one domino trips the next.
- **The epidemiological model.** This model sees accidents as related to latent failures that hide in everything from management decisions to procedures to equipment design. These "pathogens" do not normally wreak havoc unless they are activated by other factors.
- **The systemic model.** This model sees accidents as emerging from *interactions* between system components and processes, rather than failures within them. As such, accidents come from the normal workings of the system; they are a systematic by-product of people and organizations trying to pursue success with imperfect knowledge and under the pressure of other resource constraints (scarcity, competition, time limits).

A model helps you determine what things to look for. It brings some kind of order into the rubble of failure because it suggests ways in which you can explain relationships. So the accident model that you believe in—probably without knowing it—is enormously helpful.

But that model is also constraining. After all, if the model tells you to look for certain things, and look at those things in a particular way, you may do just that—at the exclusion of other things, or at the exclusion of interpreting things differently. Just consider how the different models help you think about preventing the next accident:

- In the sequence-of-events model, accidents can be prevented by taking one link from the chain; by removing one domino, or by inserting a barrier between any two dominoes.
- In the epidemiological model, accidents can be prevented by identifying and knocking out resident pathogens, or by making sure they don't get activated.
- In the systemic model, accidents can get prevented by understanding better how people and organizations normally function; how people and organizations do or do not adapt effectively to cope with complexity.

Different models are good at explaining different things. But different models also have different problems and can be misleading in varying ways. For example:

- In the sequence of events model, the choice of events considered causal to one another is subjective and always incomplete. The starting point of the sequence is an arbitrary choice too, as prior events can always be added. Also, humans often get painted as "the weakest link" in the chain (an Old View idea: unreliable people undermine safe systems).

In one explosive decompression incident, Leveson reports, a DC-10 airliner lost part of its passenger floor, and thus all the control cables that ran through it, when a cargo door opened in flight in 1972. The pilot had actually trained himself to fly the plane using only the engines because he had been concerned about exactly such a decompression-caused collapse of the floor. He was able to land the plane safely. Afterwards, he recommended that all DC-10 pilots be trained in engine-only flying techniques. The Safety Board, the regulator and a subcontractor to McDonnell Douglas (the manufacturer of the plane) all recommended changes to the design of the aircraft. McDonnell Douglas, however, attributed the incident to "human error": the baggage handler responsible for closing the cargo compartment door had not done his job properly. It was shown to be very difficult to see whether the door had actually latched, but a "human error" here represented a convenient, and cheap, event in the chain to stop at.[3]

- In the epidemiological model, searches for "latent pathogens" can quickly become pointless, as everything can be construed as a possible latent failure inside an organization. While the model is helpful for finding latent failures in the rubble after a mishap, it is more difficult to make meaningful predictions with it.

I recall talking to the safety manager of an Air Traffic Control organization. He had commissioned a study into "latent pathogens" in his organization, in order to get a better estimate of the fragility (or accident liability) of his company. The epidemiological idea should help people estimate, or predict how close an organization had come to breakdown, and help direct countermeasures to places where they could have most effect.

When the safety manager's team crossed the mark of 837 latent pathogens, it gave up and stopped counting. There was no end in sight. With a bit of fantasy, everything could potentially be a latent pathogen. What exactly was a latent pathogen? Where in practice should you look for one (or rather, not look for one)? Was a latent pathogen a hole in a layer of defense? A lack of a layer of defense? A potential interaction of one layer of defense with another?

- The systemic model explains accidents in terms of inadequate control over the interactions between components and processes. But because our systems are growing increasingly complex, such interactions are difficult to model. Also, "inadequate control" over them is not easy to define. As these interactions are most often a structural by-product of the system's normal functioning, the line between success and failure in system models is very thin.

So, next to their strengths, the different models have different drawbacks that make them inappropriate for certain uses. The Field Guide will draw variously from the three different models to help you cover different aspects of the failure you may be dealing with.

The Sequence-of-Events Model

The sequence of events model is very good for explaining the last few minutes (or seconds, or perhaps hours, depending on the time constants in your domain) before an accident, and how the events during that time could be related to the outcome. One technical problem leads to, or causes, another. All the events in the sequence are necessary and only jointly sufficient to produce the outcome failure.

The "physical cause" of the loss of the Space Shuttle Columbia in February 2003 was "a breach in the Thermal Protection System on the leading edge of the left wing. The breach was initiated by a piece of insulating foam that separated from the left bipod ramp of the External Tank and struck the wing in the vicinity of the lower half of Reinforced Carbon-Carbon panel 8 at 81.9 seconds after launch. During re-entry, this breach in the Thermal Protection System allowed superheated air to penetrate the leading-edge insulation and progressively melt the aluminum structure of the left wing, resulting in a weakening of the structure until increasing aerodynamic forces caused loss of control, failure of the wing, and breakup of the Orbiter."[4]

The sequence-of-events model can deal well with cause-effect relationships. In fact, that is what it consists of. It can certainly deal well with technical cause-effect relationships. If causes and effects are pretty obvious, and if not too many causes lead to too many effects, a sequence-of-events model can help you explain an outcome failure. This can work well for the events immediately before a mishap.

Sequence-of-events models tell you a relatively simple story of what caused an accident. You can keep it stable in your own head without too much trouble, and can communicate it clearly in your reports. You can even represent it graphically. You can also nicely point to countermeasures (removing one domino, putting a barrier between others) that will ensure that this sequence never leads to trouble again.

Which, indeed, it hardly will. Even if you did nothing. The problem with the linear thinking behind sequence-of-events models is that it is easily outrun by the real complexity of real systems. The pathway (calling it that already makes you a believer in the first kind of accident model) towards failure is seldom linear or narrow or simple. Consider the following three figures. Here is a supposedly linear pathway to what we can call an "automation surprise". Its sequence of events can be represented as in Figure 10.1.

In the case of an automation surprise of Figure 10.1, the automation controlling the path of an airliner did something the human pilots had not expected. The pilots had been asked to expedite a climb (climb faster), so they switched their autopilot to "vertical speed" mode in which the pilots can dial in a rate of climb. The aircraft couldn't meet the dialed-in target because it was too warm outside and the aircraft too heavy.

Registering that it couldn't climb any further, the automation reverted to "Altitude Hold" mode. For lack of anything better to come up with, it kept the aircraft at a particular altitude, while the speed declined further and further. At last, the aircraft stalled: it could no longer keep flying because it was going too slowly.

In the sequence-of-events model, a countermeasure could be to no longer use the vertical speed mode when asked to expedite a climb. In other words, make

What is Your Accident Model? 85

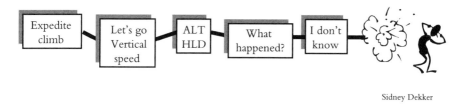

Figure 10.1 **Laying out a sequence of events, including people's assessments and actions and changes in the process itself.**

a rule, write a procedure. This puts a barrier between the request to expedite a climb and the use of the vertical speed mode. This is represented in Figure 10.2.

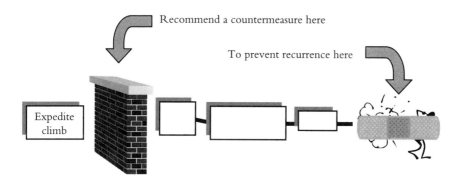

Figure 10.2 **We may believe that blocking a known pathway to failure somewhere along the way will prevent all similar mishaps.**

But mishaps have dense patterns of causes, with contributions from all corners and parts of the system, and typically depend on many subtle interactions. Putting one countermeasure in place somewhere along (what you thought was like) a linear pathway to failure may not be enough (see Figure 10.3).

Erik Hollnagel reminds us that looking for causes in a sequence of events is a bit like the drunk outside in the night, looking for his housekeys. He looks for the keys not where he lost them, but in the light of the streetlamp. That, after all, is where the light is better.

In general, it is extremely unlikely that the precise sequence of events, or confluence of factors, will repeat itself, whether you put a barrier there or not. Complex systems have many parts that interact and are tightly coupled to one another. They can generate unfamiliar, unexpected interactions that are not immediately visible or comprehensible. Just slicing through one sequence of interactions you have now understood by looking at it in hindsight, leaves the door wide open for others to develop.

I was participating in an air traffic control investigation meeting that looked into a loss of separation between two aircraft (they came too close together in flight). When the meeting got to the "probable cause", some controllers proposed that the cause be "the clearance of one of these aircraft to flight level 200"—after all, that was the cardinal mistake that "caused" the separation loss. They focused, in other words, on the last proximal act that could have avoided the incident, but didn't, and thus they labeled it "the cause".

Such focus on the proximal is silly, of course: many things went before and into that particular clearance, which itself was only one act out of a whole stream of assessments and actions, set in what turned out to be a challenging and non-routine situation. And who says it was the last act that could have avoided the incident? What if the aircraft could have seen each other visually, but didn't? In that case there would be another proximal cause: the failure of the aircraft to visually identify each other. One controller commented: "if the cause of this incident is the clearance of that aircraft to flight level 200, then the solution is to never again clear that whole airline to flight level 200".

Of course, putting the barrier "earlier" in a sequence of events can supposedly prevent a larger class of problems or failures. But as a result, the rule or procedure (or other kind of barrier) will be very context-insensitive. It may force itself into situations where it does no good at all, and where it may even do harm by restricting the creative problem-solving people or organizations to meet to get out of trouble.

In devising countermeasures, it is crucial to understand the vulnerabilities through which entire parts of a system (the tools, tasks, operational and organizational features) can contribute to system failure under different guises or conditions. This is why the "Findings as to risk" as used in the Swissair MD-11 report (see the previous chapter) are so powerful. Investigators there highlight factors through which the entire industry is exposed to problems that played a role in their accident (for example, not knowing how contamination

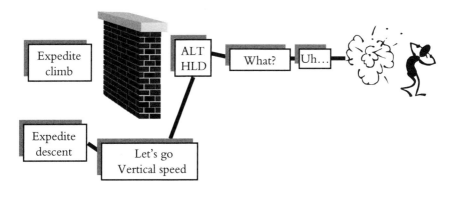

Figure 10.3 Without understanding and addressing the deeper and more subtle vulnerabilities that surround failure, we leave opportunities for recurrence open.

affects the continued airworthiness of an aircraft once it has been certified as safe to fly).

Those investigators abandoned the ambition to think about causes linearly, or to represent things graphically, or to even keep them stable in the head. They traded the illusion of linear simplicity for the complexity of real system understanding. It gave them a better explanation, and it certainly gave them a better shot at helping prevent similar accidents.

The Epidemiological Model

When it comes to the organizational factors behind the creation of an accident, the sequence-of-events model does not help you understand or prove much. For one, it does not invite you to go back a whole lot. As Jens Rasmussen reminded us, it is difficult to backtrack the causal sequence *through* a human, and human actions or inactions (at whatever level in an organization) are often seen as the legitimate stopping point (or "root cause").

The sequence-of-events model is also too linear, too direct, too narrow, too incomplete to capture organizational contributions well. For example, you can assert that production pressure "caused" people to take shortcuts. But presenting evidence for that is a whole different matter, and the sequence-of-events model leaves you empty-handed. How does the connection between production pressure and shortcuts look? With the first accident model, you can't say much about *how* one influences the other, or leave much room for other factors that may be involved.

Because of such limitations, large-scale industrial accidents during the late 1970s and 1980s inspired the creation of a new kind of accident model, the epidemiological one. It sees accidents as an effect of the combination of:

- *Active errors* or unsafe acts, committed by those on the sharp end of a system, and
- *Latent errors*, or conditions buried inside the organization that lie dormant for a long time but can be triggered in a particular set of circumstances.

The story of the escape of huge amounts of methyl isocyanate (MIC) from Union Carbide's pesticide plant in Bhopal, India, in 1984 is one of many latent failures, that combined with more active problems on a fateful night. For example, instrumentation in the process control room was inadequate: its design had not taken extreme conditions into account: meters pegged (saturated) at values far below what was actually going on inside the MIC tank. Defenses that could have stopped or mitigated the further evolution of events either did not exist or came up short. For example, none of the plant operators had ever taken any emergency procedures training. The tank refrigeration system had been shut down and was now devoid of liquid coolant; the vent gas scrubber was designed to neutralize escaping MIC gasses of quantities 200 times less and at lower temperatures than what was actually escaping; the flare tower (that would burn off escaping gas and was itself intact) had been disconnected from the MIC tanks because maintenance workers had removed a corroded pipe and never replaced it. Finally, a water curtain to contain the gas cloud could reach only 40 feet up into the air, while the MIC billowed from a hole more than 100 feet up.

The view that accidents really are the result of long-standing deficiencies that finally get activated, has turned people's attention to upstream factors, away from frontline operator "errors". The aim is to find out how those "errors" too, are a systematic product of managerial actions and organizational conditions. It encourages you to probe organizational contributions to failure and to see "human error" at the sharp end not as a cause, but an effect.

James Reason has said it most colorfully: "Rather than being the main instigators of an accident, operators tend to be the inheritors of system defects created by poor design, incorrect installation, faulty maintenance and bad management decisions. Their part is usually that of adding the final garnish to a lethal brew whose ingredients have already been long in the cooking."[5]

What is Your Accident Model?

Epidemiological models allow you to think in terms other than causal series. It offers you the possibility of seeing more complex connections between various factors, and it certainly helps you in your search for organizational issues behind the creation of an accident.

The epidemiological model has nonetheless had a hard time departing from the sequential idea. Look at Figure 10.4, the signature image of the model: it is a sequence. The accident trajectory through the defense layers is a straight line. First it penetrates one barrier, then the next, and so on. The problem is that there is little evidence from accidents that this is indeed how failure propagates. Moreover, the decomposition into linear layers of defense may be misleading our understanding of what makes systems risky or safe.

Even though it may linearlize and oversimplify, the epidemiological model has been helpful in portraying the resulting imperfect organizational structure that let an accident happen. You can imagine the anteceding organization as a series of porous defense layers. But how did that structure come about? Epidemiological models do not specify much about the processes that create the holes in the layers of defense. They also do not explain how active and latent failures interact (they simply, magically do sometimes), nor do they tell you how or why the holes line up. This is left up to your imagination, and indeed,

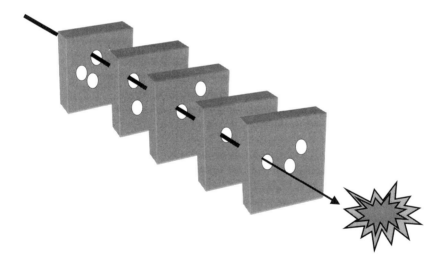

Figure 10.4 The "Swiss Cheese" analogy. Latent and active failures are represented as holes in the layers of defense. These need to line up for an accident to happen (after Reason, 1990).

it has inspired a generation of investigators to think more creatively about the organizational precursors of accidents.

The Systemic Accident Model

An important limitation of epidemiological models is that they rely heavily on "failures" up and down an organizational ladder ("poor design, incorrect installation, faulty maintenance and bad management decisions").

The problem is that if these actions were seen as "failures" at the time, people would have done something about them, or done something different. The epidemiological model does little to help you understand how people at all levels in an organization, from their perspective, could have seen these conditions or decisions as normal or rational. Remember, the point of the New View is to understand why it made sense for people to do what they did. Otherwise you won't be able to prevent other people from doing the same thing. The point of the New View is to resist labeling people's design as "poor" or their management decisions as "bad". That just replaces one error with another.

If an accident model relies on failures to explain failure, it cannot shed any light on the conditions or actions that were responsible for an accident, but that people themselves, on the inside, saw as perfectly normal at the time. The task, for you, is to understand why a decision made sense. Not to judge it "bad" in hindsight. For this, epidemiological models put you on the wrong footing.

While the "physical cause" of the Columbia Space Shuttle accident can be traced by a sequence-of-events model, the organizational processes behind it can't. Even epidemiological models fall short. The challenge is to explain why people gradually come to see deviant behavior or system performance as normal or acceptable, not judge it deviant from the outside and hindsight.

The accident focused attention on the maintenance work that was done on the Shuttle's external fuel tank, revealing the differential pressures of having to be safe and getting the job done (better, but also faster and cheaper). A mechanic working for the contractor, whose task it was to apply the insulating foam to the external fuel tank, testified that it took just a couple of weeks to learn how to get the job done, thereby pleasing upper management and meeting production schedules. An older worker soon showed him how he could mix the base chemicals of the foam in a cup and brush it over scratches and gouges in the insulation, without reporting the repair.

The mechanic found himself doing this hundreds of times, each time without filling out the required paperwork. Scratches and gouges that were brushed over with the mixture from the cup basically did not exist as far as the organization was concerned. And those that did not exist could not hold up the production schedule for the external fuel tanks. Inspectors often did not check. A company program that once had paid workers hundreds of dollars for finding defects had been watered down, virtually inverted by incentives for getting the job done now.

Telling you to look for holes and failures makes you forget that these are normal people doing normal work. You will never understand why it is normal to them; why what they did made sense to them. This limitation of epidemiological models has given rise to systemic accident models.

The systems approach focuses on the whole, not the parts (like the other two accident models do). It does not help you much to just focus on human errors, for example, or on an equipment failure, without taking into account the socio-technical system that helped shape the conditions for people's performance and the design, testing and fielding of that equipment.

The interesting properties of systems (the ones that give rise to system accidents) can only be studied and understood when you treat them in their entirety. Taking them apart, and examining individual components or processes, grossly distorts the analysis results to the point that nothing of value can be learned or predicted. System models build on two fundamental ideas:[6]

- *Emergence*: Safety is an emergent property that arises when system components and processes interact with one another and their environment. Safety can be determined only by seeing how parts or processes fit together within a larger system;
- *Control* imposes constraints on the degrees of freedom (e.g. through procedures, design requirements) of components, so as to control their interaction. Such control is based on prior (and possibly false) ideas about how components and processes interact. Control can be imperfect and even erode over time.

System accidents result not from component failures, but from inadequate control or enforcement of safety-related constraints on the development, design and operation of the system.[7] This allows the possibility that people and organizations may have thought that their control and enforcement of safety-related constraints was adequate—based on their knowledge, goals and focus of attention at the time. But systems are not static designs; they are dynamic processes, continually adapting to achieve their goals in a changing environment. What was adequate control before may now have eroded constraints and pushed the system closer to the edge, operating with smaller margins.

So in contrast to the other two models, systemic accident models do not rely on a component breaking or a human erring. In fact, they do not have to rely on anything "going wrong", or anything being out of the ordinary. They will help you realize that even though an accident happened, nothing really went wrong—in the sense that nothing that happened was out of the ordinary. They will help you explain accidents by a confluence of a number of things, just on the border of the ordinary.[8]

Systemic models see accidents as a by-product of the *normal* functioning of the system, not the result of something breaking down or failing inside of that system. In that sense, there is not much difference (if at all) between studying a successful or a failed system, since the organizational processes that lead to both are very similar. Systemic accident models, in other words, are hardly "accident models". They are models of a system's or organization's normal functioning.

An added advantage of systemic accident models is that they can deal with non-linear interactions. They do not need to rely on linear cause-effect relationships, as the other two models do, to explain how factors interact or relate to one another. This means that they can stay much closer to the real complexity behind system success and failure. It also means that they, as models, are more complex.

The accident model you believe in, or choose as foundation for your understanding of human error, determines in part what kind of data you need to pursue. The next chapter (Chapter 11) is about getting human factors data, and Chapter 12 proposes how you can use time as an organizing principle to get some order in (and information out of) your data.

Notes

1 Hollnagel, E. (2004). *Barriers and accident prevention.* Aldershot, UK: Ashgate Publishing Co.
2 Leveson, N.G. (2002). *A new approach to system safety engineering.* Cambridge, MA: MIT Aeronautics and Astronautics.
3 Ibid., p. 17.
4 Columbia Accident Investigation Board (2003). *Report Volume 1*, August 2003. Washington, DC: US Government Printing Office.
5 Reason, J.T. (1990). *Human Error.* Cambridge, UK: Cambridge University Press, p. 173.
6 Leveson, N.G. (2002). *Op. cit.*, p. 43.
7 Ibid.
8 Hollnagel, E., Woods, D.D., and Leveson, N.G. (2006). *Resilience engineering: Concepts and precepts.* Aldershot, UK: Ashgate Publishing Co.

11 Human Factors Data

Before you can begin to re-assemble the puzzle of human performance, you need data. What data do you need, and where do you get it? The first hunch is to say that you need everything you can get. As system models remind you (see the previous chapter), the people whose performance you are trying to understand did not work in a vacuum. They performed by touching almost every aspect of the system around them, and were touched by it in return.

For example, in the most proximal focus, a nurse administered a tenfold drug dose to a baby patient who subsequently died. So you need data about drug labeling and packaging. You also need data about for example fatigue, drug side effects, about scheduling pressures, about task saturation and workload, cognitive fixation, about external distractions, about care-giver-to-patient-ratios relative to time of day or night, about drug administration and double-checking procedures, about physician supervision, hospital organization, regulatory oversight, and probably much more.

Human factors is not just about humans, just like human error is not just about humans. It is about how features of people's tools and tasks and working environment systematically influence human performance. So you need to gather data about all the features that are relevant to the event at hand. This chapter looks at two direct sources of human factors data that can give you a good start; good indications of where to begin looking further:

- Debriefings of participants;
- Recordings of performance parameters.

Both have their own advantages and pitfalls. And they are only the beginning. You may have to do much archeological work inside the organization to find out about production pressures, goal conflicts, procedural constraints—things that help you understand not only what people did, but also why. More about those issues will follow in later chapters.

Debriefings of Participants

What seems like a good idea—ask the people involved in the mishap themselves—also carries a great potential for distortion. This is not because operators necessarily have a desire to bend the truth when asked about their contribution to failure. In fact, experience shows that participants are interested in finding out what went wrong and why, which generally makes them forthright about their actions and assessments. Rather, problems arise because of the inherent features of human memory:

- Human memory does not function like a videotape that can be rewound and played again;
- Human memory is a highly complex, interconnected network of impressions. It quickly becomes impossible to separate actual events and cues that were observed from later inputs;
- Human memory tends to order and structure events more than they were; it makes events and stories more linear and plausible.

Gary Klein has spent many years refining methods of debriefing people after incidents: firefighters, pilots, nurses, and so forth. Insights from these methods are valuable to share with investigators of human error mishaps here.[1]

The aim of a debriefing

Debriefings of mishap participants are intended primarily to help reconstruct the situation that surrounded people at the time and to get their point of view on that situation. Some investigations may have access to a re-play of how the world (for example: cockpit instruments, radar displays, process control panel) looked during the sequence of events, which may seem like a wonderful tool. It must be used with caution, however, in order to avoid memory distortions. Klein proposes the following debriefing order:

1. First have participants tell the story from their point of view, without presenting them with any replays that supposedly "refresh their memory" but would actually distort it;
2. Then tell the story back to them as investigator. This is an investment in common ground, to check whether you understand the story as the participants understood it;
3. If you had not done so already, identify (together with participants) the critical junctures in a sequence of events;

4   Progressively probe and rebuild how the world looked to people on the inside of the situation at each juncture. Here it is appropriate to show a re-play (if available) to fill the gaps that may still exist, or to show the difference between data that were available to people and data that were actually observed by them.

At each juncture in the sequence of events (if that is how you want to structure this part of the accident story), you want to get to know:

- Which cues were observed? (What did he or she notice/see or did not notice what he or she had expected to notice?)
- What knowledge was used to deal with the situation? Did participants have any experience with similar situations that was useful in dealing with this one?
- What expectations did participants have about how things were going to develop, and what options did they think they have to influence the course of events?
- How did other influences (operational or organizational) help determine how they interpreted the situation and how they would act?

Some of Klein's questions to ask

Here are some questions Gary Klein and his researchers typically ask to find out how the situation looked to people on the inside at each of the critical junctures:

Cues	What were you seeing?
	What were you focusing on?
	What were you expecting to happen?
Interpretation	If you had to describe the situation to your colleague at that point, what would you have told?
Errors	What mistakes (for example in interpretation) were likely at this point?
Previous experience/ knowledge	Were you reminded of any previous experience?
	Did this situation fit a standard scenario?
	Were you trained to deal with this situation?
	Were there any rules that applied clearly here?
	Did you rely on other sources of knowledge to tell you what to do?

Goals	What goals governed your actions at the time?
	Were there conflicts or trade-offs to make between goals?
	Was there time pressure?
Taking action	How did you judge you could influence the course of events?
	Did you discuss or mentally imagine a number of options or did you know straight away what to do?
Outcome	Did the outcome fit your expectation?
	Did you have to update your assessment of the situation?

Debriefings need not follow such a tightly scripted set of questions, of course, as the relevance of questions depends on the event under investigation. Also, the questions can come across to participants as too conceptual (e.g. the one about goal conflicts) to make any sense. You may need to reformulate them in the language of the domain.

Dealing with disagreements and inconsistencies in debriefings

It is not uncommon that operators change their story, even if slightly, when they are debriefed on multiple occasions. Also, different participants who were caught up in the same events may come with a different take on things. It is difficult to rule out the role of advocacy here—people can be interested in preserving an image of their own contribution to the events that may contradict earlier findings or statements from others. How should you deal with this as investigator? This mostly depends on the circumstances. But here is some generic guidance:

- Make the disagreements and inconsistencies, if any, explicit in your account of the event.
- If later statements from the same people contradict earlier ones, choose which version you want to rely on for your analysis and make explicit why.
- Most importantly, see disagreements and inconsistencies not as impairments of your investigation, but as additional human factors data for it. Mostly, such discontinuities can point you towards goal conflicts that played a role in the event, and may likely play a role again.

Recordings of Performance Data

One thing that human error investigations are almost never short of is wishes for more recorded data, and novel ideas and proposals for capturing more performance data. This is especially the case when mishap participants are no longer available for debriefing. Advances in recording what people did have been enormous—there has been a succession of recording materials and strategies, data transfer technologies; everything up to proposals for permanently mounted video cameras in cockpits and other critical workplaces. In aviation, the electronic footprint that a professional pilot leaves during every flight is huge, thanks to monitoring systems now installed in almost every airliner.

Getting these data, however, is only one side of the problem. Our ability to make sense of these data; to reconstruct how people contributed to an unfolding sequence of events, has not kept pace with our growing technical ability to register traces of their behavior. The issue that gets buried easily in people's enthusiasm for new data technologies is that recordings of human behavior—whether through voice (for example Cockpit Voice Recorders) or process parameters (for example Flight Data Recorders)—are never the real or complete behavior.

Recordings represent partial data traces: small, letterbox-sized windows onto assessments and actions that all were part of a larger picture. Human behavior in rich, unfolding settings is much more than the data trace it leaves behind. Data traces point beyond themselves, to a world that was unfolding around the people at the time, to tasks, goals, perceptions, intentions, and thoughts that have since evaporated. The burden is on investigators to combine what people did with what happened around them, but various problems conspire against their ability to do so:

Conventional restrictions

Investigations may be formally restricted in how they can couple recorded data traces to the world (e.g. instrument indications, automation mode settings) that was unfolding around the people who left those traces behind. Conventions and rules on investigations may prescribe how only those data that can be factually established may be analyzed in the search for cause (this is, for example, the case in aviation). Such provisions leave a voice or data recording as only factual, decontextualized and impoverished footprint of human performance.

Lack of automation traces

In many domains this problem is compounded by the fact that today's recordings may not capture important automation-related traces—precisely the data of immediate importance to the problem-solving environment in which many people today carry out their jobs. Much operational human work has shifted from direct control of a process to the management and supervision of a suite of automated systems, and accident sequences frequently start with small problems in human-machine interaction.

Not recording relevant traces at the intersection between people and technology represents a large gap in our ability to understand human contributions to system failure. For example, flight data recorders in many automated airliners do not track which navigation beacons were selected by the pilots, what automation mode control panel selections on airspeed, heading, altitude and vertical speed were made, or what was shown on either of the pilots' moving map displays. This makes it difficult to understand how and why certain lateral or vertical navigational decisions were made, something that can hamper investigations into CFIT accidents (Controlled Flight Into Terrain—an important category of aircraft mishaps).

The Problem with Human Factors Data

One problem with understanding human error is the seeming lack of data. You may think you need access to certain process or performance parameters to get an understanding not only of what people did, but why. Solutions to this lack may be technically feasible, but socially unpalatable (e.g. video cameras in workplaces), and it actually remains questionable whether these technical solutions would capture data at the right resolution or from the right angles.

This means that to find out about critical process parameters (for instance, what really was shown on that left operator's display?) you will have to rely on interpolation. You must build evidence for the missing parameter from other data traces that you *do* have access to. For example, there may be an utterance by one of the operators that refers to the display ("but it shows that it's to the left…" or something to that effect) which gives you enough clues when combined with other data or knowledge about their tasks and goals.

Recognize that data is not something absolute. There is not a finite amount of data that you could gather about a human error mishap and then think you have it all. Data about human error is infinite, and you will often have to reconstruct certain data from other data, cross-linking and bridging between different sources in order to arrive at what you want to know.

This can take you into some new problems. For example, investigations may need to make a distinction between factual data and analysis. So where is the border between these two if you start to derive or infer certain data from other data? It all depends on what you can factually establish and how factually you establish it. If there is structure behind your inferences—in other words, if you can show what you did and why you concluded what you concluded—it may be quite acceptable to present well-derived data as factual evidence.

Note

1   See: Klein, G. (1998). *Sources of power: How people make decisions*. Cambridge, MA: MIT Press.

12 Build a Timeline

Time is a powerful organizing principle, especially if you want to understand human activities in an event-driven domain. Event-driven means that the pace of activities is not (entirely) under control of the humans who operate the process. Things are happening in the process itself, or to it, that determine the tempo of for example people's situation assessment and decision making.

In event-driven domains, people's work ebbs and flows in sync with the processes they manage. The amount of work, the demands it poses, the pressures it puts on people—all of this varies over time. The kinds of work that people engage in (e.g. troubleshooting, assessing, deciding, rechecking) also vary over time, in concert with demands posed by the process. If you want to begin to under-stand human error (e.g. why people seem to have missed things, or decided things that, in hindsight, do not seem to make sense) a good starting point is to build a timeline. This timeline, however, needs to be of sufficient resolution to reveal underlying processes that may be responsible for the "errors". In many attempts to understand human error, timelines are of poor quality. Not because the data to build a good one weren't available, but because people did not realize the importance of a sufficiently detailed timeline for gaining insight into human performance issues.

Your timeline should be of maximum possible resolution to reveal human performance issues

Remember, at all times, what you are trying to do. In order to understand other people's assessments and actions, you must try to attain the perspective of the people who were there at the time. Their decisions were based on what they saw on the inside of the tunnel—not on what you happen to know today.

This chapter will first show you three ways of building a timeline around people's communication, each of increasing resolution:

- What was said or done (low resolution timeline)
- What was said or done when (medium resolution timeline)
- What was said or done when and how (high resolution timeline)

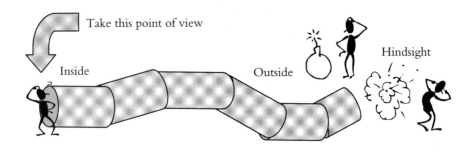

Figure 12.1 Remember at all times to try to see the unfolding world from the point of view of people inside the situation—not from the outside or from hindsight.

As the grain of analysis gets finer, you will see that deeper worlds of human performance start to open up for you. But just building a timeline of communication may not be enough. The chapter will then continue with timelines that may better show you how people's assessments and actions are coupled to what was going on in the process they managed and monitored.

Two issues before you go on

First, the datatraces in this chapter may be impossible for you to construct, simply because you do not have access to such data (e.g. voice recordings of what operators said). However, other ways of getting time-dependent data may be available. You could have event logs, for example, that operators are required to keep. Or you may get machine read-outs of settings and processing that went on. These could allow you to infer what kinds of human activities were happening when (which things were set when, which in turn points to particular human assessments or decisions). Remember, the monitored process (e.g. operation on a patient, journey of a ship, flight of an airplane, handling alarms in process control) evolves in parallel with human activities to manage that process, so one usually allows you to infer what was (or must have been) going on in the other.

While the kind of data you have access to can restrict the resolution of your analysis, that should not keep you from striving to tweak as much information out of your datatrace as possible. For that, the sections below may still help.

The second problem is the beginning of your timeline. Despite its strengths as an analytic tool, a timeline imports all the difficulties and limitations of the sequence-of-events model (see Chapter 10). What should the beginning of a sequence of events be? There is inherent difficulty in deciding what counts as the beginning (especially the beginning—the end of a sequence of events often speaks for itself). Since there is no such thing as a root cause (remember Chapter 9), there is technically no such thing as the beginning of a mishap.

Yet if you want to understand human error, you need to start somewhere. Beginning with the people who were closest in space and time to the eventual mishap is often the default choice. This may be fine, as long as you realize that the whole point of laying out their timeline is to understand *why* they did what they did. Beginning with what they did is only the first step for digging into the background that explains why. For that, you may have to go much higher or deeper, and much further back. You may have to derive inspiration from accident models other than the sequence-of-events one.

For example, the beginning of a voice recording may be where investigative activities start for real, and the end of the recording may be where they end. Or the beginning is contained in the typical 72-hour or 24-hour histories of what a particular practitioner did and did not do (play tennis, sleep well, wake up early, etc.) before embarking on a fatal journey or operation. Of course even these markers are arbitrary. This is bad if you don't make clear what your reasons are for picking them as your starting point.

Someone else can always say that another decision or action preceded the one that you marked as your starting point. Such disagreement serves as a reminder for you of what to take into account when analyzing the decision or action you have chosen as your beginning. What went on before that? Whatever your choice of beginning, make it explicit. From there you can reach back into history, or over into surrounding circumstances, and seek explanations for the decision or action that, according to you, set the sequence of events in motion.

Was the pilot's acceptance of a runway change the trigger of trouble? Or was it the air traffic controller's dilemma of having too many aircraft converge on the airport at the same time—something that necessitated the runway change? Or was it the aircraft's entry into this controller's airspace?

Making clear where you start, and explaining this choice, is essential for a structured, well-engineered human error investigation.

Low-Resolution Communication Timeline

Look at the following interaction between two people involved in managing an event-driven process. This could be any process and any two people managing it (e.g. an anaesthesiologist and her assistant discussing machine-assisted drug delivery to a patient during surgery; two pilots discussing settings during an approach into an airport; two operators of a powerplant talking about intervening in their process; an officer on a ship's bridge discussing a course change with a subordinate).

The interaction is fictitious and short and deliberately generic, so that you don't get lost in the details of any particular domain at this stage. The point is to see how human performance issues are, or are not, brought to the fore by the way you use time to structure your data. Suppose that the interaction below led up to a mishap later on, because the people involved missed whatever it is they had to set (drug delivery rate, minimum safe altitude, steam pressure level, ship heading). Now you need to find out how features of the interaction may have played a role in creating the conditions for that mishap. Here is the low-resolution timeline:

15:19:44	P1	We'll go down (to) one forty three
15:20:23	P1	You set it?
15:20:32	P1	Uh, you can put the steps in there too if you don't mind
15:20:36	P2	Yeah, I was planning ★★.
15:20:37	P1	But you only need to put the steps in, ah, below the lowest safe
15:20:41	P2	Okay it's set

15:19:44 = Local Time (hours:minutes:seconds)
P1 = Person 1
P2 = Person 2
() = Editorial insertion
★★ = Unintelligible word

At first sight, not much seems wrong with the timeline above. The discussion between the two people is captured over time. Person 1 is talking more than person 2. Person 2 also seems to have a more leading role in managing the process, taking initiative and evidently directing person 2 to do certain things in certain ways. Person 2 ends up making the setting and acknowledges as much (at 15:20:41).

Many people involved in understanding human error could be content to stop here. After all, the trace here allows them to draw conclusions about, for example, the quality of the communication between person 1 and 2. There is no closed loop, to mention one obvious issue: person two says "Okay it's set", but does not close the loop on the target setting called for at 15:19:44, nor on the intermediate steps that apparently also could, or should, be set. People could draw a conclusion about "miscommunication" being responsible for the mishap. Or, even more vaguely, a breakdown of effective crew coordination. But what, really, is the evidence for a breakdown in effective crew coordination, based on the trace above? The persons involved could have been coordinating in other, nonverbal ways, for example, and closing communication loops that way.

Another aspect of the low-resolution timeline is that it actually does not use time as an organizing principle at all. Instead, it uses excerpts of communication as its organizating principle. The only occasions where you get a little window on the unfolding process is when somebody says something. And then this format hangs time (the starting times of those excerpts) on the side of it. Basically, this timeline says what was said (and perhaps what was done). As to when it was said, this timeline provides a mere ordinal account of what came before what. A lot is lost when you represent your data that way. Let us move to a trace of higher resolution and see what it reveals.

Medium-Resolution Communication Timeline

If you want to learn about human performance, you really have to take time as organizing principle seriously. The reason is that phenomena such as taskload, workload management, stress, fatigue, distractions, or problem escalation are essentially meaningless if it weren't for *time*. They would not happen in real life if it weren't for the role time played in them. And you cannot go back to a datatrace and discover them (or argue for them) either if you do not give time a more prominent role.

Increasing the resolution of the timeline, merely by representing the data at a finer grain of analysis, could help you reveal other things about, in this case, the coordination between the two persons here. These things easily get lost in the low-resolution representation. Or at least they do not jump out, demanding explanation or clarification. Here is the same trace, at a higher resolution:

15:19:44   P1   We'll go down
15:19:45   (to) one
15:19:46
15:19:47
15:19:48   forty three
15:19:49
15:19:50
15:19:51
15:19:52
15:19:53
15:19:54
15:19:55
15:19:56
15:19:57
15:19:58
15:19:59
15:20:00
15:20:01
15:20:02
15:20:03
15:20:04
15:20:05
15:20:06
15:20:07
15:20:08
15:20:09
15:20:10
15:20:11
15:20:12
15:20:13
15:20:14
15:20:15
15:20:16
15:20:17

15:20:18		
15:20:19		
15:20:20		
15:20:21		
15:20:22		
15:20:23	P1	You set it?
15:20:24		
15:20:25		
15:20:26		
15:20:27		
15:20:28		
15:20:29		
15:20:30		
15:20:31		
15:20:32	P1	Uh,
15:20:33		you can put the steps in
15:20:34		there too if you don't
15:20:35		mind
15:20:36		P2 Yeah, I was plan-
15:20:37	P1	But you only need to ning
15:20:38		put the steps in
15:20:39		ah
15:20:40		below the lowest
15:20:41		safe P2 Okay it's set

Now the ebbs and flows in activity become visible. Now you can start to see places of more talk and less talk. Also, by marking both the beginning *and* the end of utterances (as opposed to only the beginning), you can see areas of overlap, of a crunching together of communication, of one person breaking off the other. None of this would have been as compellingly visible if you had left your data as in the low-resolution representation.

The medium representation allows you to start asking new questions. Take the period of silence after the first remark from person 1. It takes more than half a minute until person 1 speaks again. The nature of the question at 15:20:23 could indicate that nothing has been set to go down to 143 yet during that half minute. This empty space in the transcript demands an explanation. What was going on instead? Was person 2 distracted or occupied with another task, as was person 1? Person 1 may have been doing something else too, then gave person 2 the assignment for a 143 setting, and then got occupied again with the other duties. So the period of silence may not mark a period of low workload (perhaps the contrary).

Once you have worked this out further, you can begin to offer an explanation for why person 1 raises the issue again at 15:20:23. It could begin to reveal something about surprise (why hasn't it been done yet?). At least person 1 now may believe that person 2 needs more careful coaching or monitoring in the execution of that task. As about 10 seconds go by, and person 2 is (presumably) making the setting, person 1 could be supervising what person 2 is doing, with the suggestion 15:20:32 as expression and evidence of this.

This suggestion, in turn, seems to be superfluous as far as person 2 is concerned, as person 2 was already "planning to" do so (presumably). Person 1, in turn, may not be interested in what person 2 was planning to do, as the coaching and suggesting continue, in finer detail now. Until person 2 breaks it off while person 2 is still talking, by saying "Okay it's set", leaving the "it" unspecified. Rather than a communication loop not being fully closed, this excerpt from person 2 could serve a message about roles and the dislike of how they are being played out. It could mark impatience with having been assigned to do the task in the first place and then be micromanaged in doing it. "Okay it's set" may really be saying "Leave me alone now".

Of course, these are all leaps of faith. These are leaps from the data trace into conclusions about what—psychologically or socially—went on between these people and the process they managed. Claims about what people may have understood or felt, must be based on, and demonstrated in analyses of the transcribed data[1]. The more detailed transcript gives you an ability to start arguing around these possible interpretations. But you still need more data, and analysis, to start substantiating them. The low-resolution transcript never even really brought the questions into being. Your data could have gone underanalyzed.

Yet, despite the level of detail, you may still not feel entirely comfortable with the conclusions. Closing this gap, from data to conclusion, in a tracable way, is critical if you want your human performance explanation to be taken seriously.

High-Resolution Communication Timeline

Maurice Nevile, working with accident investigators, has led the way in applying a technique called *conversation analysis* to voice transcripts. This section is based on his work.[2] Conversation analysis uses an even finer-grained notation for systematically respresenting all kinds of aspects of talk (and non-talk activities). In comparison to the medium-resolution timeline, a timeline constructed using conversation analysis reveals even more about *how* things are said and done. It can make maximally visible how people themselves develop and understand their respective contributions to interaction and to the work required to operate their system. For example, a timeline using conversation analysis shows:

- The order of talk: how and when do participants switch roles as listener and speaker and how much of each role is played by the different participants;
- How much silence there is and when this happens;
- What overlap is there where different people talk simultaneously;
- Features of the manner of talk (stretching sounds, pitching up or down, slowing down talk or speeding it up, louder or softer talk);
- So-called tokens, such as "oh", "um" and "ah".

These features are not just interesting in themselves. They all mean something, based on elaborate theories of social interaction. They can point to the nature of people's interaction in managing a process, and how that interaction may have contributed to, or detracted from, managing an unfolding situation safely.

This is what the exchange between our two people could look like using notation from conversation analysis:

```
        (1.2)
P1      We'll go down (to) one (2.5) forty (.) three.
        (34.3)
P1      You set it?
        (8.1)
P1      Uh (0.9) you can put the (.) <steps in there too> °if you don't
        mi::nd°
        (0.2)
P2      Yeah, I was [pla::°nning ( )°]
P1                  [But you only:.] need to put the steps in (0.8) ah
        (1.1) below the lowest [safe, ]
P2                             [>Okay] it's s::et<
        (1.0)
```

And this is what the codes and notations mean:

(1.2)	pauses in seconds and tens of seconds
(.)	micro pause (shorter than two-tenths of a second)
>set<	faster than surrounding talk
<set>	slower than surrounding talk
set	louder than surrounding talk
°set°	quieter than surrounding talk
set,	flat or slightly rising pitch at the end
set.	falling pitch at the end
set?	rising pitch at the end
se::t	falling pitch within word
se::t	rising pitch within word
set:.	falling, then rising pitch
set:.	rising, then falling pitch
[set]	overlapping with other talk (also set in [ ])
( )	talk that could not be transcribed
(set)	doubt about the talk transcribed
(set/sit)	doubt about the talk/possible alternative

There are, of course, diverging opinions about what each of these things may mean for a particular utterance or the quality of people's interaction. But with a high-resolution notation like the one above, you can begin to pick things out that the other timelines did not allow to the same extent. For instance:

- The pause in of 2.5 seconds could mean that P1 is figuring the setting out while uttering the suggestion to P2, or may even be looking it up somewhere while speaking;
- The emphasis on "set it" in P1's next utterance;
- Slowing down the "steps in there too" could point to an increasing hesitancy to interfere during that moment. Perhaps P1 is realizing that the suggestion is superfluous, as P2 is already doing so;
- An increasing hesitancy to interfere on part of P1 could be confirmed by the following quieter "if you don't mind". Also, the rising pitch in the word "mind" could make it more mildly suggestive, pleasantly inquisitive;
- The overlap between P2's increasingly silent, trailing-off "planning" and the loud start of P1's next suggestive sentence ("But you only...") as well as the rising/falling pitch in "only" could point to yet another reversion in how P1's experiences his/her role relative to P2. P2 may be seen to require more forceful, concrete guidance in making the setting;

- The subsequent overlap between the last part of P1's latest suggestion ("… the lowest safe") and P2's very quick and initially loud "Okay it's set" could mean that P2 is getting unhappy with the increasingly detailed prodding by P1. The falling pitch in "set" would confirm that P2 now wants the matter to be closed.

Taken individually, such features do not represent "errors" on the part of the people interacting here. But together, and in how they relate to each other, these features can be creating a context in which successful task performance is increasingly unlikely. As you will see in the next chapter, this type of analysis gives you the opportunity to provide detailed, concrete evidence for a phenomenon such as "loss of effective crew resource management".

Connecting Behavior and Process

Just analyzing people's conversation may not tell you all you need to know about how they managed an ongoing process. What people do, and what happens in their monitored process, is often highly intertwined. The three timelines above do not couple process and behavior. You may need other representations of people's unfolding work context. For example, if you want to understand the nature of the half-minute delay in making the setting (see above), it may not be enough to know what people said when and how. You also want to answer these questions:

- What was going on in the process at that time?
- What other tasks would people have plausibly been involved in simultaneously, if at all?

What was going on around people at the time?

The reconstruction of mindset begins with the unfolding situation in which the mind found itself. You can begin to understand what people may have been thinking, looking at, if you rebuild their evolving situation.

Tracing relevant process parameters over time, and coupling them to a record of communication, is a good approach. This is the first step toward coupling behavior and situation—toward putting the observed behavior back into the situation that produced and accompanied it. The point is to marry the two as an inextricable, causal dance-à-deux. You can even build a picture that shows these connections. Here is how you can do that:

- Find out how process parameters were changing over time, both as a result of human influences and of the process moving along—make a trace of changing pressures, ratios, settings, quantities, modes, rates, and so forth.
- Find out how the values of these parameters were available to people—dials, displays, knobs that pointed certain ways, sounds, mode annunciations, alarms, warnings. Their availability does not mean people actually observed them: you have to make that distinction (see later).
- Decide which—of all the parameters—counted as a stimulus for the behavior under investigation, and which did not. Which of these indications or parameters, and how they evolved over time, were instrumental in influencing the behavior in your mishap sequence?

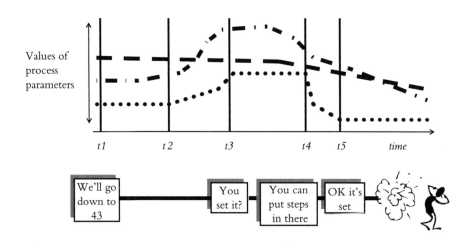

Figure 12.2 Connecting critical process parameters to the sequence of people's assessments and actions and other junctures.

You can capture the connection between people's assessments and actions nicely by building a layout that combines time and space. This can show the route to the outcome not only over time but also how it meandered through a landscape (the approach path to an airport or harbor, for example). Be sure to coordinate the scales: if actions and assessments are separated by time as well as space, indicate so clearly on your representation of the sequence of events by spacing them apart—to scale. As said before, a presentation of how a situation unfolded over time (and through space) is the basis for a credible human factors analysis.

Tasks over time

You may now have laid out all relevant parameters, but what did people actually notice? What did they understand their situation to be? The answer lies in part in people's goals. People were in the situation to get a job done; to achieve particular aims. The goals that drive people's behavior at the time will tell you something about the kinds and numbers of tasks they would be trying to accomplish, and thereby the things they would, or would not, be looking for or at.

Finding what tasks people were working on does not need to be difficult. It often connects directly to how the process was unfolding around them. To identify what job people were trying to accomplish, ask yourself the following questions:

- What is canonical, or normal at this time in the operation? Jobs relate in systematic ways to stages in a process. You can find these relationships out from your own knowledge or from that of (other) expert operators.
- What was happening in the managed process? Starting from your record of parameters from the previous picture (Figure 12.2), you can see how systems were set or inputs were made. These changes obviously connect to the tasks people were carrying out.
- What were other people in the operating environment doing? People who work together on common goals often divide the necessary tasks among them in predictable or complementary ways. There may be standard role divisions, for example between pilot flying and pilot not-flying, that specify the tasks for each. What one operator was doing may give some hints about what the other operator was doing.

If you find that pictures speak more clearly than text, you can create a graphical representation of the major tasks over time, and, if necessary, who was carrying out what. This picture can also give you a good impression of the taskload in the sequence of events. See Figure 12.3 for an example. Once you can see what people may have been busy with, you may begin to understand what they were looking at and why.

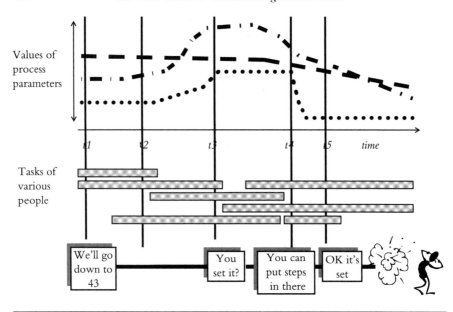

Figure 12.3 Laying out the various (overlapping) tasks that people were accomplishing during an unfolding situation.

How do you Identify Events in your Data?

So how do you identify the events that can serve as a thread through the presentation of your data? As shown in the timelines in the first half of this chapter, people's utterances may be strong markers for identifying events. But they may not be the only ones:

- Other things may have happened in the process itself, without people talking about it or commenting on it;
- Or people may have been doing things to the process, or apparently been drawing conclusions about it, without necessarily verbalizing those;
- You can also infer people's interpretations from their subsequent actions in managing the process, rather than from what they say.

In a study on automation surprises, researchers were interested to find out if people noticed the loss of a particular signal in a critical phase of operations. The automation showed whether it received the signal or not, but this indication easily got hidden beneath other indications. It was easy to miss. In this case, researchers did not draw conclusions about people's awareness of the signal loss on the basis of what they said (or didn't say), but rather on the basis of their next actions. If people continued to operate the process as if nothing had happened, it was likely that they had not seen the loss of signal. Debriefings afterward confirmed this.

So if you want more event anchors than just people's utterances, here are some areas to focus on:

- Places where people did something or (with your knowledge of hindsight) could have done something to influence the direction of events;
- Places where the process did something or could have done something to influence the direction of events—whether as a result of human inputs or not;
- Short episodes where either people or the processes they managed contributed critically to the direction of events and/or the outcome that resulted;
- As a rule, what people did and what their processes did is highly interconnected. Finding events in one can or should lead you to events in the other.

Here are some examples of typical events, whether accompanied by talk or not:

- **Decisions** can be obvious events, particularly when they are made in the open and talked about. A point where people chose not to decide anything is still a decision and still an event.
- **Shifts in behavior.** There may be points where people realized that the situation was different from what they believed it to be previously. You can see this either in their remarks or their actions. These shifts are markers where later you want to look for the evidence that people may have used to come to a different realization.
- **Actions to influence the process** may come from people's own intentions. Depending on the kind of data that your domain records or provides, evidence for these actions may not be found in the actions themselves, but in process changes that follow from them. As a clue for a later step, such actions also form a nice little window on people's understanding of the situation at that time.
- **Changes in the process.** Any significant change in the process that people manage must serve as an event. Not all changes in a process managed by

people actually come from people. In fact, increasing automation in a variety of workplaces has led to the potential for autonomous process changes almost everywhere—for example:

- Automatic shut-down sequences or other interventions;
- Alarms that go off because a parameter crossed a threshold;
- Uncommanded mode changes;
- Autonomous recovery from undesirable states or configurations.

Yet even if they are autonomous, these process changes do not happen in a vacuum. They always point to human behavior around them; behavior that preceded it and behavior that followed it. People may have helped to get the process into a configuration where autonomous changes were triggered. And when changes happen, people notice them or not; people respond to them or not. Such actions, or the lack of them, again give you a strong clue about people's knowledge and current understanding.

The events that were no events

Human decisions, actions and assessments can also be less obvious. For example, people seem to decide, in the face of evidence to the contrary, to not change their course of action; to continue with their plan as it is. With your hindsight, you may see that people had opportunities to recover from their misunderstanding of the situation, but missed the cues, or misinterpreted them.

These "decisions" to continue, these opportunities to revise, may look like clear candidates for events to you. And they are. But they are events only in hindsight. To the people caught up in the unfolding situation, there probably was no compelling reason to re-assess their situation or decide against anything. Or else they would have. They were doing what they were doing because they thought they were right; given their understanding of the situation; their pressures. The challenge for you becomes to understand how this was not an event to the people you were investigating. How their "decision" to continue was nothing more than continuous behavior—reinforced by their current understanding of the situation, confirmed by the cues they were focusing on, and reaffirmed by their expectations of how things would develop.

Notes

1. Nevile, M., and Walker, M.B. (2005). *A context for error. Using conversation analysis to represent and analyse recorded voice data (Aviation Research Report B2005/0108)*. Canberra, Australia: Australian Transport Safety Bureau.
2. Ibid.

13 Leave a Trace

"A spokesman for the Kennedy family has declined to comment on reports that a federal investigation has concluded that pilot error caused the plane crash that killed John F. Kennedy Jr., his wife and his sister-in-law. The National Transportation Safety Board is expected to finish its report on last year's crash and release it in the next several weeks. Rather than use the words 'pilot error', however, the safety board will probably attribute the cause to Kennedy's becoming 'spatially disoriented', which is when a pilot loses track of the plane's position in the sky."[1]

Underspecified Labels

"Human error" as explanation for accidents has become increasingly unsatisfying. As mentioned earlier, there is always an organizational world that lays the groundwork for the "errors", and an operational one that allows them to spin into larger trouble.

We also know there is a psychological world behind the errors—to do with people's attention, perception, decision making, and so forth. Human factors has produced or loaned a number of terms that try to capture such phenomena. Labels like complacency, situation awareness, crew resource management, shared mental models, stress, workload, or non-compliance are such common currency today that nobody really dares to ask what they actually mean. The labels are assumed to speak for themselves; to be inherently meaningful. They get used freely as causes to explain failure. For example:

- "The crew lost situation awareness and effective crew resource management (CRM)" (which is why they crashed);
- "High workload led to a stressful situation" (which is why they got into this incident);
- "It is essential in the battle against complacency that crews retain their situation awareness" (otherwise they keep missing those warning signals);
- "Non-compliance with procedures is the single largest cause of human error and failure" (so people should just follow the rules).

The question is: are labels such as complacency or situation awareness much better than the label "human error"? In one sense you would think they are. They provide some specification; they appear to give some kind of reasons behind the behavior; they provide an idea of the sort of circumstances and manner in which the error manifested itself.

But if they are used as quoted above, they do not differ from the verdict "human error" they were meant to replace. These labels actually all conclude that human error—by different names—was the cause:

- Loss of CRM is one name for human error—the failure to invest in common ground, to coordinate operationally significant data among crewmembers;
- Loss of situation awareness is another name for human error—the failure to notice things that in hindsight turned out to be critical;
- Complacency is also a name for human error—the failure to recognize the gravity of a situation or to follow procedures or standards of good practice;
- Non-compliance is also a name for human error—the failure to stick with standard procedures that would keep the job safe.

Dealing with the illusion of explanation

Human factors risks falling into the trap of citing "human error" by any other name. Just like "human error", labels like the ones above also hide what really went on, and instead simply judge people for what they did not do (follow the rules, coordinate with one another, notice a signal, etc.). Rather than explaining why people did what they did, these labels say "human error" over and over again. They get you nowhere. As shown in earlier chapters, judging people is easy. Explaining why their assessments and actions made sense is hard. The labels above may give you the illusion of explanation, but they are really judgments in disguise. Saying that other people lost situation awareness, for example, is really saying that you now know more about their situation than they did back then, and then you call it their error. This of course explains nothing.

Don't Make Leaps of Faith

Here is why labels like "complacency" or "loss of situation awareness" must not be mistaken for deeper insight into human factors issues. And why they certainly should not be mistaken for satisfactory explanations of human error.

When you say something like "high workload led to a loss of crew resource management" (which then led to the mishap), you are really making two leaps of faith:

- The first leap you make is from the data or factual information (e.g. what people said or did at the time) and the labels you put on that data. According to you, this data is evidence for high workload and a loss of crew resource management. But have you proved your point at all? Have you left a trace that convincingly couples the factual information to your large-label conclusion? You see, the problem is that the factual information may, to somebody else, point to an entirely different condition. And they could be just as right—if all you do is make large-label claims because the data "looks like" people got busy and didn't coordinate their assessments and actions anymore.
- The second leap of faith you make is from the large labels to the outcome (the mishap). How is it that the psychological conditions you invoked "caused" the mishap? Most efforts to understand human error leave this entirely to the imagination. For example, "loss of effective crew resource management" may be cited in the probable causes or conclusions. But how exactly did the behaviors that constituted its loss contribute to the outcome of the sequence of events?

As shown in Figure 13.1, you end up hiding all kinds of interesting things by pasting a large label over your factual data. You can only hope it will serve as a meaningful explanation of what went wrong and why. But it won't. And if you really want to understand human error, posting a large label is no substitute for doing hard analytic work.

The use of large terms in findings and explanations of human error may be seen as a rite of passage into psychological phenomena. That is, for a human error explanation to be taken seriously, it should contain its dose of situation awarenesses and stresses and workloads. But in the rush to come across like somebody who really understands human error, you may forget that you can't just jump from the specifics in your evidence to a large label that seems to cover it all. You need to explain something in between; you need to leave a trace. Otherwise other people will get completely lost and will have no idea whether you are right or not. This is what we call "folk science", or "folk psychology" and human factors explanations can be full of it.

- Folk models substitute one big term for another instead of defining the big term by breaking it down into more little ones (in science we call this

122   The Field Guide to Understanding Human Error

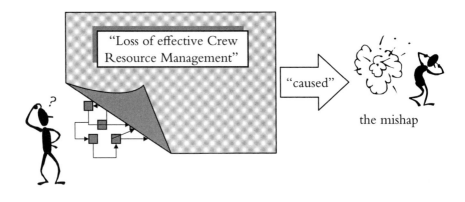

Figure 13.1   The interesting cognitive and coordinative dynamics take place *beneath* the large psychological label. The label itself explains nothing.

decomposition, or deconstruction). So instead of human error, you would simply say "loss of CRM". But you still don't explain anything.
- Folk models are difficult to prove wrong, because they do not have a definition in terms of smaller components, that are observable in people's real behavior. Folk models may seem glib; they offer popular, but not necessarily helpful, characterizations of difficult phenomena.
- Folk models easily lead to overgeneralization. Before you know it, you may see "complacency" and "loss of situation awareness" everywhere. This is possible because the concepts are so ill-defined. You are not bound to particular definitions, so you may interpret the concepts any way you like.

Take as an example an automation-related accident that occurred when concepts of situation awareness or automation-induced complacency did not yet exist—in 1973. The issue was an aircraft on approach in rapidly changing weather conditions that was equipped with a slightly deficient "flight director" (a device on the central instrument showing the pilot where to go, based on an unseen variety of sensory inputs), and which the captain of the airplane in question distrusted. The airplane struck a seawall bounding Boston's Logan airport about a kilometer short of the runway and slightly to the side of it, killing all 89 people on board. In its comment on the crash, the transport safety board explained how an accumulation of discrepancies, none critical in themselves, can rapidly deteriorate, without positive flight management, into a high-risk situation. The first officer, who was flying, was preoccupied with the information presented by his flight director systems, to the detriment of his attention to altitude, heading and airspeed control.

Today, both automation-induced complacency on part of the first officer and a loss of situation awareness on part of the entire crew would most likely be cited under the causes of this crash. (Actually, that the same set of empirical phenomena can comfortably be grouped under either label (complacency or loss of situation awareness) is additional testimony to the undifferentiated and underspecified nature of these human factors concepts). These "explanations" [complacency, loss of situation awareness] were obviously not necessary in 1973 to deal with this accident. The analysis left us instead with more detailed, more falsifiable, and more traceable assertions that linked features of the situation (e.g. an accumulation of discrepancies) with measureable or demonstrable aspects of human performance (diversion of attention to the flight director versus other sources of data). The decrease in falsifiability represented by complacency and situation awareness as hypothetical contenders in explaining this crash is really the inverse of scientific progress. Promise derives from being better than what went before—which these models, and the way in which they would be used, are not.

From Context to Concept

Look at Figure 13.2. You see two streams of events there:

- The one below is the one with your factual data. This is what you understand went on at the time. People looked here, did that, moved over there, turned a knob, threw a switch, said this or that to another person involved, and so forth. This is the sort of thing you can end up with as a result of constructing a timeline (see the previous chapter). This is the level of context, because it describes events in a context-dependent or domain-dependent language.
- The one on top is the stream of concepts that you think explains all of those data from your timeline. This may be one concept, such as loss of crew resource management, or multiple, for example if there are also issues of automation surprises and plan continuation (see the next chapter). These may partially overlap, depending on how the situation developed. This is the level of concept, as it describes the same events in a conceptual language.

The challenge is to "prove" that your concept fits your data. That it explains or covers your data. You can't just ask people to take your word for it. You have to make your concept(s) carry the explanation of your data convincingly.

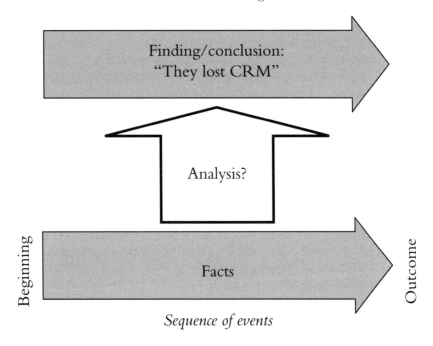

Figure 13.2 Don't make a leap of faith, from your facts to a big label that you think explains those facts. Leave an analytic trace that shows how you got to your conclusion.

That means demonstrably. If you don't prove this by leaving some kind of analytic trace, anybody else can come in and assert that the data from the timeline are actually related to, say, bad system design, not loss of crew resource management. And if that person does not offer much proof either, all that you are left with is two competing claims.

Investigations into human error do this a lot. The leaving of a trace is not yet common practice. And for the lack of it, a lot of breath is wasted on discussions about whose claim carries the most explanatory weight. The biggest problem is that the claims cannot really be argued with (and arbitration is also impossible) because none shows how they got there.

So how should you do the hard analytic work to leave a trace:

- Between your factual data and your conclusion about them; and
- Between the conclusion and the outcome of the mishap?

In other words, how can you become confident (and convince other people) that the things that went on can be explained by, for example, a loss of crew resource management? And how can you be confident that the loss of crew resource management helped produce the outcome? Those are two critical questions. Without looking at them in detail, your efforts to understand error (and explain error to other people) will be less powerful and may even become useless.

There are basically two ways in which you can start to build an analytic trace:

- **From the bottom up**. Here you pull together critical features of your data that form the constitutive features of a phenomenon (you will see later what conversation analysis sees as constituting a loss of effective crew resource management). You can avoid cherry picking only if you come with a substantive theory that explains how these bits hang together. Otherwise you will once again have no analytic trace. And people will be left to take your word for it.

You can't, for example, establish the role of fatigue in an accident just because your datatrace has a yawn here and somebody mentioning that he or she is tired there. A yawn and a remark may be hints at fatigue, but they are not sufficient to prove its existence. You will have more persuasion work to do.

- **From the top down**. Here you begin not with your data, but with a definition for a large concept (e.g. loss of effective CRM) from the literature. With this in your hand, you descend into your data to look if you can find evidence for the things specified by that definition. This works only if the definition indeed breaks up the phenomenon into smaller, measurable or observable bits. Otherwise it will simply be a folk model.

In practice, building an analytic trace is often a give-and-take between top-down and bottom-up strategies. Beginning bottom up, you may recognize things that remind you of some phenomenon. You then go to the top, where you search the literature for that phenomenon and find its constitutive, or defining, features. Then you descend back into your data and start looking around for those.

You have to keep an open mind in this, of course. Your first impressions may turn out false. Don't stretch your data too far just because you want to stick with your first choice of phenomenon, for it may quickly lose credibility.

A DC-9 airliner got caught in windshear close to an airport. This is a phenomenon where the airspeed rapidly dwindles because of a sudden change from headwind to tailwind. In trying to explain why the pilot pushed the nose of the DC-9 down rather than pull it up and supposedly out of the situation, the NTSB offered that "the circumstances during the last minute of flight strongly suggested that the captain, upon losing his visual cues instantaneously when the airplane encountered the heavy rain, could have experienced a form of spatial disorientation. The disorientation may have led him to believe that the aircraft was climbing at an excessively high rate and that the pitch attitude should be lowered to prevent an aerodynamic stall. Additionally, when the airplane encountered the heavy rain, the flightcrew would have lost their outside horizon visual reference. Also, it may not have been possible for the captain to regain situational focus on the primary flight instruments because he was performing other tasks. Further, because the flightcrew was initiating the missed approach, which involved a large increase in engine thrust, a pronounced increase in pitch attitude, and a banked turn to the right, the crew would have been exposed to significant linear and angular accelerative forces. These forces could have stimulated the flightcrew's vestibular and proprioceptive sensory systems and produced a form of spatial disorientation known as somatogravic illusion."

This may be true, but a huge number of facts would need to be checked in order to really build an empirical base for the explanation above. How did the crew lose their outside visual reference, and what role did it play? Was the pilot performing "other tasks"? How significant was the linear and angular force? And so forth. A simpler explanation would be that the pilot of US1016 felt the bottom fall from under his airplane, because the airspeed was bleeding off very quickly. A plausible response would have been to push the nose down to regain airspeed.

In science this has been called "Occam's razor": the explanation that makes the smallest number of assumptions is the preferable one, if anything because it is more sustainable or will cost less effort to prove. In other words, the simpler or shorter, the better. This still does not justify folk modeling. Saying the accident was due to "a loss of crew resource management", for example, may sound simple. But it actually makes so many (untestable) assumptions that it would never make the cut of Occam's razor.

These strategies for leaving an analytic trace apply to any of the patterns of failure you think explains your data. You will find more such patterns in the next chapter.

Leave a Trace I: Bottom Up

One way to begin with a trace is to go back to your data and do more systematic analysis on that. As you have seen in the previous chapter, conversation analysis is one technique to do so. Recall the interaction between person 1 and person 2:

```
        (1.2)
P1      We'll go down (to) one (2.5) forty (.) three.
        (34.3)
P1      You set it?
        (8.1)
P1      Uh (0.9) you can put the (.) <steps in there too> °if you don't
        mi::nd°
        (0.2)
P2      Yeah, I was [pla::°nning ( )°]
P1                  [But you only:.] need to put the steps in (0.8) ah
        (1.1) below the lowest [safe, ]
P2                             [>Okay] it's s::et<
        (1.0)
```

With help from the notation of conversation analysis,[2] several features stand out:

- Turn-taking in the interaction is not divided equally. Person 1 does much more of the talking then person 2. This may of course be naturally related to the roles that the two people have. But it is something that you could sort out further. What would other pairs of people do in this situation?
- Suggestions for what to do often come from P1, with P2 answering either with a reply that indicates the suggestion may have been superfluous or even disruptive, or with silence. When one conversation partner regularly witholds talk, or opts out, or replies only in a clipped fashion, you can learn something meaningful about people's collaboration. Silence here (e.g. after "one forty three") is not a matter of nobody speaking, but of P2 not replying.
- There is overlapping talk in multiple places. Significantly, when there is overlapping talk, the other person still has talk of substance to utter (for example about what P2 was "planning" to do, or what P1 meant with the "lowest safe"). The interruptions cut such talk of substance short, perhaps eroding a joint conduct and understanding of the developing process and how to manage it.

Other features that do not stand out clearly from the interchange above include "repair". We often engage in repair of our own talk, for example when we say something unclearly or when we detect some other communicative problem. Then we go back and say it more clearly, or say it differently. It isn't usual for people to often repair each other's talk, so if this does happen a lot, you may begin to suspect a problem in the interaction between them.

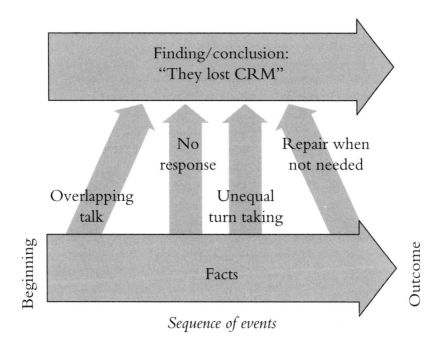

Figure 13.3 Leaving a trace bottom up. Overlapping talk, no response when one is expected, unequal turns at talking and offering repair of somebody else's talk when none is needed together could point to a "loss of effective CRM".

Taken together, these features can create what Maurice Nevile calls an *interactional context for error* or a *collaborative construction of error*. Because of the nature of the interaction, errors may become more likely, and their detection and recovery become less likely. Rather than just saying "a loss of effective crew resource management", the sort of analysis above allows you to put some meat on your argument. It allows you to leave a trace.

Leave a Trace II: Top Down

The other way to leave a trace is to start with the large concept, rather than with your data that instantiates that concept. Consulting the human factors literature can help you break down the large phenomenon that you suspect is at play. By breaking it down into more "measurable" or identifiable bits, you can begin your search for concrete evidence in your data.

Take, for example, from Judith Orasanu at NASA what effective crew resource management is about:

- Shared understanding of the situation, the nature of the problem, the cause of the problem, the meaning of available cues, and what is likely to happen in the future, with or without action by the team members;
- Shared understanding of the goal or desired outcome;
- Shared understanding of the solution strategy: what will be done, by whom, when, and why?[3]

The literature is always a good place to start looking for definitions of complex phenomena (such as crew resource management). Judith Orasanu's definition of team resource management (or shared problem models as she also calls it) is so good because it "deconstructs" the big label of CRM. It breaks the label down into more specific components, for which you can actually seek evidence in your facts.

Such deconstruction, or breaking down a large concept into smaller, more specific ones, is absolutely critical for drawing credible conclusions. It is hard to find, or prove, evidence for simply a "loss of CRM". It is much easier to find, to prove and to defend, that you found evidence for a lack of common understanding of, for example, the cause of a problem. Evidence for this can be found in what people do or do not say to one another, for instance.

The other advantage of deconstructing a large concept is that other people can actually choose to agree or disagree with your definition. If you don't define, for example, "situation awareness", but simply assert that the operator in question "lost situation awareness", then other people cannot even agree or disagree – because they would not know with what. Everybody else simply has to take your word for it, which is not a good way to understand human error. The point in breaking down your definition is not to be right. Rather, it is about offering other people the opportunity to show where you got your conclusions from.

If you define loss of CRM, like above, you give other people specific information with which they can agree or disagree—both conceptually

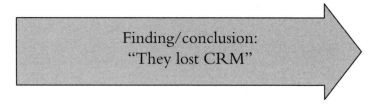

Define "Crew Resource Management" and state criteria for when in is lost

Locate evidence for this in your facts

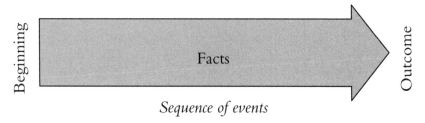

Figure 13.4 **The way to bridge the gap between facts and conclusions (about those facts) from the top down is to find a definition in the literature for the phenomenon you suspect is at play, and start looking in your facts for evidence of it.**

(whether a loss of CRM is these things) and whether there is evidence for it in this specific case. For the latter, other people can have different opinions about what counts as evidence for a shar ed understanding of a solution strategy, for instance. For the former, they can claim that a loss of CRM has more to do with the tone, the nature, the mechanics of the interaction than with its substance (just look at the previous section). And that is OK too. Both allow you to verify and strengthen (or even refute and replace) arguments rather than repeat large labels.

Notice how breaking down the phenomenon takes discussions a notch lower into detail. Discussions around evidence for whether people understood a solution strategy similarly or not, can make much more sense than arguments about evidence for a "loss of CRM". Your whole investigation becomes a lot more credible. And by leaving such an analytic trace, you will find yourself contributing much more to the understanding of human error.

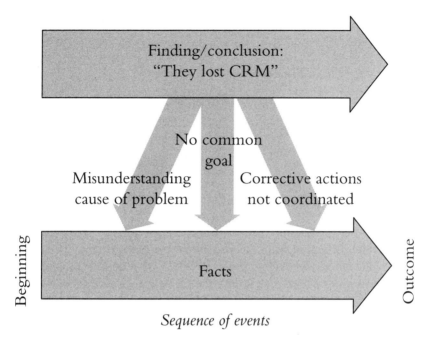

Figure 13.5 Leaving a trace top-down. Using a definition for "loss of effective CRM" that lists misunderstanding the problem, no common goal and uncoordinated corrective actions, you can find evidence for that in your facts.

Folk Models in Human Factors

To be sure, the literature does not always help you here. Human factors research itself is not immune to folk modeling. For example, if you look for a definition of "complacency" that breaks down that phenomenon into more "measurable" or identifiable bits, well, stop looking. You probably won't really find it. Here is what a good sample of the literature has equated complacency with:[4]

- Overconfidence
- Self-satisfaction
- Trait that can lead to a reduced awareness of danger
- State of confidence plus contentment
- Low index of suspicion
- Unjustified assumption of satisfactory system state

- Loss situation awareness, and unpreparedness to react in timely manner when system fails.

These are not definitions, these are substitutions. One big label replaces another. It does not explain or specify anything in any greater detail. For example, self-satisfaction takes the place of complacency and is assumed to speak for itself. Just imagine: you ask, 'what do you mean with complacency?' and you get the answer 'well …, self-satisfaction'. You then are supposed to go, 'Of course! Now I understand!' But you don't understand. Because this is no explanation, it is a substitution. Can you look for "self-satisfaction" in your data any better or more convincingly than you can look for "complacency"? Hardly. There is no explanation (or breakdown) of a psychological mechanism that makes self-satisfaction emerge. You are none the wiser.

If the literature can't provide ways to define and identify the phenomenon in question, how are you to argue for its existence in real situations? This is impossible. So this is where you should start getting suspicious. Which is precisely what Nevile Moray does when it comes to complacency.[5] Complacency is a mere judgment from outside the tunnel. You now know more than the person on the inside, and point out a difference between what the person in the situation sampled or focused on versus what he should have sampled or focused on. The difference is "complacency" (or "overconfidence", or any of the other labels).

But even such a judgment is impossible to make. To prove that somebody was complacent, Moray argues, you must show that this person sampled a variable less often than is optimal, given the dynamics of what is going on in the system at that time. But it is very difficult to rigorously define the optimal sampling rate in supervisory monitoring. Just try this for an application in your own world.

Moreover, you cannot claim that somebody was complacent because he or she missed a piece of data (that you, in hindsight, find important). Complacency, after all, is about under-sampling or defective monitoring (which is impossible to establish because you can't define the optimal). It is not about whether people detected available signals that you deem observable in hindsight. Observability (or detectability) is different from, and much more complex than availability. It is a function of signal-to-noise ratio and somebody's response criterion (as in, when do I have enough evidence to do something), not of sampling strategy.

Complacency, then, will remain a folk model because it is based on folk science. It will remain a folk model because it is nothing more than an after-the-fact judgment that says the person should have done something different from what he or she did. It is not a psychological explanation of anything. As soon as you ask for hard data (what is, for instance, the optimal sampling that you compare your data with?), the entire concept evaporates into fantasy, as any of it is impossible to establish seriously.

Notes

1   *International Herald Tribune,* June 24–25, 2000.
2   Nevile, M., and Walker, M.B. (2005). A context for error. Using conversation analysis to represent and analyse recorded voice data (Aviation Research Report B2005/0108). Canberra, Australia: Australian Transport Safety Bureau.
3   Orasanu, J., Martin, L., and Davison, J. (in press). Cognitive and contextual factors in aviation accidents: Decision errors. In E. Salas and G. Klein (eds) *Applications of naturalistic decision making*. Mahwah, NJ: Lawrence Erlbaum Associates.
4   Dekker, S.W.A., and Hollnagel, E. (2004). Human factors and folk models. *Journal of Cognition, Technology and Work*, 6, 79–86.
5   Moray, N., and Inagaki, T. (2001). Attention and complacency. *Theoretical Issues in Ergonomics Science*, 4, 354–365.

14 So What Went Wrong?

So, what went wrong? This chapter introduces a few typical mechanisms by which human performance can get into trouble. They can apply to all kinds of settings where people perform safety-critical work. Here they are:

- Cognitive fixation
- Plan continuation
- Stress
- Fatigue
- Buggy or inert knowledge
- New technology and computerization
- Automation surprises
- Procedural adaptations.

This chapter can offer you no more than a selection and a flavor of these mechanisms. The research literature underneath most of these phenomena is huge. Remember, you want to avoid the folk modeling that is so common in efforts to understand human error. This chapter asks *and* allows you to leave a trace, to show how you get to your conclusion from your data.

Cognitive Fixation

One of the most vexing issues in understanding human error is how other people did not seem to see the situation for what it really was (or turned out to be, which you now know thanks to your hindsight). Human factors has even coined labels to deal with this issue, saying for example how "they lost situation awareness". This, of course, is nothing but Old View banter. It says "human error" all over again. As said in Chapter 3, such characterizations place you firmly outside the sequence of events. They reveal your complete ignorance about why people on the inside actually did what they did.

Psychologically, a "loss of situation awareness" is nonsensical too. Because if you lose situation awareness, what replaces it? There is no such thing as a cognitive vacuum. People always have an idea about where they are, what the world around them looks like. Of course, these ideas may be incomplete or even wrong (compared to what you now know the world to have been), but saying that this is a "loss" of anything is useless. Rather, the point is to understand how this picture may have made sense, or felt complete or accurate, to people at the time.

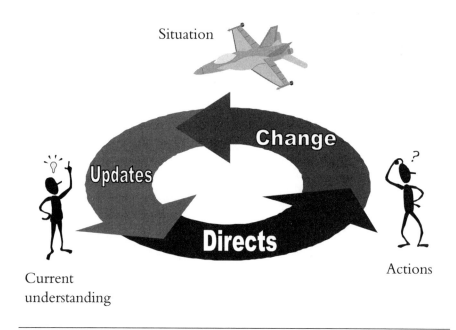

Figure 14.1 We make assessments about the world, which update our current understanding. This directs our actions in the world, which change what the world looks like, which in turn updates our understanding, and so forth (figure is modeled on Neisser's perceptual cycle).

Sensemaking is ongoing. People's actions and assessments of what is going on are deeply intertwined. By doing something, people generate more information about the world. This in turn helps them decide which actions to take next. Such realizations are not new in human factors. The basic idea of a cyclical coupling between action and assessment was propagated in the 1970s by Ulrich Neisser (see Figure 13.1).[1] The dynamics of sensemaking in unfolding situations, however, can create interesting side effects.

In 2001, the pilots of a twin-engined airliner flying from Toronto to Lisbon, with 293 passengers and 13 crew on board, noticed a fuel imbalance. One wing held more fuel than the other, which was strange: the left and right engines should consume about as much fuel from their respective tanks. The pilots reviewed engine indications and found no anomalies. Fuel flow to both engines was the same. They could think of no in-flight event or condition responsible for a fuel loss from one wingtank. They even sent a crewmember to visually check whether there was fuel leaking from the wing, but due to darkness none was seen. Handbooks and checklists did not help in resolving the mystery. In fact, the fuel quantity indications, calculations of fuel loss rates, as well as computer predictions of fuel remaining on arrival were truly "unbelievable". They made no sense. The crew concluded that they had an indication problem, but decided to divert to the Azores (Islands in the Atlantic Ocean) because of low fuel predictions in any case.

About forty-five minutes after the first fuel imbalance caution, the right engine stopped. The diversion to the Azores now became a very good idea. The idea of an indication error collapsed: this was for real. Not much later, the left engine stopped as well (the pilots earlier had crossfed fuel from one wingtank to the other). The pilots managed to glide the powerless airliner over a distance of 65 nautical miles to the Azores and land safely at Lajes Airport there. Later investigation turned up that a fuel hose in the right engine had come undone inflight because of rubbing against a non-approved part for that engine. Large quantities of fuel had been dumped overboard.

A case like the one above emphasizes the following about sensemaking, especially when people are faced with an unfamiliar or unexpected problem:

- These are situations without a well-formulated diagnosis of the problem. Handbooks and checklists are of little help.
- People have to make provisional assessments of what is going on based on partial and uncertain data.
- People's situation assessment and corrective actions are tightly interwoven. One constrains and informs the other.
- Taking action simplifies the diagnostic problem. It commits people to a particular interpretation.
- The side effect of taking action is that people then build an explanation that justifies their action. This explanation may persist and can get transformed into an assumption that is then taken for granted.

Characteristic of cognitive fixation is that the immediate problem-solving context biases people in some direction (e.g. "this is an indication problem"). From an emerging mass of uncertain, incomplete and contradictory data, people have to come up with a plausible explanation. A preliminary interpretation allows them to settle on at least a plausible explanation that covers the data observed. But it can activate certain kinds of knowledge and trouble-shooting activities at the expense of others.

The crew of the airliner above did not believe there was a fuel leak. These factors supported their interpretation of the situation:

- *The combination of the suddenness and the magnitude of the indicated fuel loss were outside anything that could reasonably be explained by a real fuel loss.*
- *There had been earlier indications of an anomaly in oil cooling. This had created uncertainty about the nature of the problem (and whether it was fuel related at all).*
- *There was no warning or caution from the airliner's computer systems warning them of a severe problem.*
- *No other indication of an engine problem was discovered: fuel flow parameters were normal.*
- *Some information, such as the cabin crew confirming that there were no visible signs of a leak, confirmed their interpretation of an indication anomaly.*

In addition, the pilots had never before experienced a fuel leak or similar events. They had not been trained for this situation. They had neither the knowledge basis nor compelling data on which to build a plausible explanation of a real fuel leak.

People who are cognitively fixated hold on to an assessment of a situation while new evidence about that situation comes in. The assessment may actually be wrong in the first place (as in the fuel leak case), with more and more evidence contradicting the assessment coming in over time. Alternatively, the assessment, while initially right, can grow increasingly at odds with how the situation is really developing.

Whether or not to abandon an initial interpretation is not about people's motivation. They are very interested in getting it right—in understanding what is going on. Instead, it is about a cognitive balancing act. When trying to understand and simultaneously manage a dynamic, uncertain situation:

- Should you change your explanation of what is going on with every new piece of data that comes in? This is called "thematic vagabonding", a jumping

around from explanation to explanation, driven by the loudest or latest indication or alarm. No coherent picture of what is going on can emerge.
- Or should you keep your explanation stable despite newly emerging data that could suggest other plausible scenarios? Not revising your assessment (cognitive fixation) can lead to an obsolete understanding.

There is no right or wrong here. Only hindsight can show you whether people should have abandoned one explanation in favor of another, or should have finally settled on a stable interpretation instead of just pursuing the latest clue. To understand human performance from the point of view of people inside the situation, you have to acknowledge the existence and importance of their balancing act. Thematic vagabonding and getting cognitively locked up are opposite extremes, created by uncertain, dynamic situations in which we ask people to solve difficult, unclear problems.

Dynamic fault management

Another aspect of managing such problems is that people have to commit cognitive resources to solving them *while* maintaining process integrity. This is called dynamic fault management, and is typical for event-driven domains. People need to keep the aircraft flying (or the patient breathing) while figuring out what is going wrong. In other words, activities associated with troubleshooting and correction come on top of normal work. At the same time, not doing troubleshooting or corrective activities well may actually challenge the integrity of the entire process (the flight may end up in the ocean, the patient may die).

In the fuel leak case described above, the dual demands of maintaining process integrity while diagnosing the problem, increased crew workload. Their work included:

- *Flying the aircraft, including navigating it to a new destination and the coordinations and coordinating with air traffic control and the airline;*
- *Diagnosing the fuel fault, which entailed looking up information in handbooks, requesting assistance from cabin crew, calling up the maintenance center, acquiring new data from computer systems and all the assessments and decisions these things triggered or called for;*
- *Managing the engine failures and preparing the aircraft for a possible ditching.*

Plan Continuation

What if the situation actually does change while people are working on a problem? This happens a lot in dynamic domains too. While on approach to an airport, for example, weather conditions may deteriorate, but crews continue anyway. While working on a patient, a problem occurs that seems to call for one type of intervention (e.g. more stimulant intravenous drugs), which eventually becomes or aggravates the source of the problem.

Sticking with the original plan while a situation has actually changed and calls for a different plan is what Judith Orasanu calls "plan continuation".[2] Plan continuation means sticking to an original plan while the changing situation actually calls for a different plan. As with cognitive fixation, situational dynamics and the continuous emergence of incomplete, uncertain evidence play a role:

- Early cues that suggest the initial plan is correct are usually very strong and unambiguous. This helps lock people into a continuation of their plan.
- Later cues that suggest the plan should be abandoned are typically fewer, more ambiguous and not as strong. Conditions may deteriorate gradually. These cues, even while people see them and acknowledge them, often do not succeed in pulling people into a different direction.

The challenge is to understand why it made sense to people to continue with their original plan. Which cues did they rely on, and why? When cues suggesting that the plan should be changed are weak or ambiguous, it is not difficult to predict where people's trade-off will go if abandoning the plan is somehow costly. Diverting with an airliner, for example, or going from laparoscopic to open surgery, entails multiple organizational and economic consequences. People need a lot of convincing evidence to justify changing their plan in these cases. This evidence may typically not be compelling until you have hindsight.

Stress and fatigue can compound plan continuation. These factors typically make it more difficult for people to entertain multiple hypotheses about a problem or project a situation into the future by mentally simulating the effects of various decision alternatives.

Stress

Stress has long been an important term, especially where people carry out dynamic, complex and safety-critical work. It may be easy to assert that people got stressed; that things got out of hand because of it. But this does not mean or explain very much. Psychologists still debate whether stress is a feature of a situation, the mental result of a situation, or a physiological and psychological coping strategy that allows us to deal with a demanding situation. This complicates the use of stress in any causal statement, because what produced what?

Demand-resource mismatch

Most theories today suggest that stress is triggered when people perceive a mismatch between the demands of the situation, and the resources they can muster to deal with those demands. What you can do on the basis of your data is make an inventory of the demands in a situation, and the resources that people had available to cope with these demands—and how each of these were perceived by people from the inside (e.g. Table 14.1).

Table 14.1 Finding a mismatch between problem demands and coping resources can help you make arguments about stress and workload more specific. Remember that people feel stress when they perceive a mismatch.

Problem demands:	Coping resources:
Ill-structured problems	Experience with similar problems
Highly dynamic circumstances: things changing quickly over time	Other people contributing to assessments of what is going on
Uncertainty about what is going on or about possible outcomes	Knowledge or training to deal with the circumstances
Interactions with other people that generate more investment than return (in terms of offloading)	Other people to off-load tasks or help solve problems
Organizational constraints and pressures	Organizational awareness of such pressures and constraints
Conflicts between goals	Guidance about goal priorities
High stakes associated with outcome	Knowledge there is an envelope of pathways to a safe outcome
Time pressure	Workload management skills

In studies of stress and workload, a mismatch between demands and resources typically means different things for different kinds of operators. In a marine patrol aircraft, for example, people in the back are concerned with dropping sonobuoys (to detect submarines) out of the aircraft. The more sonobuoys in a certain amount of time, the more workload, the more stress. People in the front of the aircraft were instead concerned with more strategic questions. For them, the number of things to do had little bearing on their experience of stress and workload. They would feel stressed, however, if their model of the situation did not match reality, or if it had fallen behind actual circumstances.

Interestingly, a small perceived mismatch between demands and resources may often lead to more stress than a large mismatch. In other words, people may experience most stress when they have the idea that they can deal with the demands, but only just—by mustering that extra bit of resources. This may happen, for example, after a change in plans. The new plan is not unmanageable (e.g. flying to a different runway), but requires more resources to be put in place than was originally counted on.

Tunneling and regression

One of the reported consequences of stress is tunneling—the tendency to see an increasingly narrow portion of one's operating environment. This is generally interpreted as a shortcoming; as something dysfunctional that marks less capable operators. Another consequence that has been noted is regression—the tendency to revert to earlier learned routines even if not entirely appropriate to the current situation.

But you can actually see both tunneling and regression as strategies in themselves; as contributions from the human that are meant to deal with high demands (lots to pay attention to and keep track of) and limited resources (limited time too look around; limited mental workspace to integrate and deal with diverse and rapidly changing data). Tunneling (sometimes called "fixation", especially when people lock onto one explanation of the world around them) comes from the human strength to form a stable, robust idea of a shifting world with multiple threads that compete for attention and where evidence may be uncertain and incomplete. The threads that get people's attention may indeed be a limited set, and may consist of the threat (e.g. a system failure) rather than the process (e.g. flying the aircraft).

In highly dynamic and complex situations, it would seem that tunneling is an (involuntary) strategy that allows people to track and stay ahead of a limited number of threads out of a host of potential ones. Similarly, regression to earlier learned routines frees up mental resources: people do not have to match current perceptions with consciously finding out what to do each time anew.

Distortion of time perception under stress

In most work, time is important. People's perception of how much time has passed can have significant consequences for the monitored process. How long has the patient been without a clear airway? How long has the aircraft been tracking towards a beacon? As other tasks pile up and need attention, the perception of how much time has passed can become distorted. People may think that it was only thirty seconds, while in reality five minutes went by.

Perception of time passing is only loosely correlated to actual time. Variations in how we perceive time probably occur in all attention-demanding situations. People's perception of time can get distorted under high workload and stress because fewer mental resources are available to keep track of time. Attention to time-giving cues (both internally and from the environment) is minimized, and no accurate track of time passed is built up in memory.

The effects show up when events in the monitored process outpace people's perception of time-passing. For example, while people were working on other attention-demanding tasks that they felt took only a few seconds, the patient may have been without an airway for too long; the aircraft may already have passed the beacon. The issue, once again, is not about people's motivation. It is about an autonomous redistribution of cognitive resources to the tasks that are deemed most relevant or important.

Fatigue

Fatigue is a common condition, especially in work that cuts across normal waking periods or even timezones. The effects of fatigue on safety-critical work are actually difficult to measure and quantify, as they are so confounded (they may well be the effect of something else, or of a combination of other factors).

Fatigue itself can actually be difficult to pin down. Using people's self-reports or their judgments of colleagues is not very reliable. Fatigue actually impairs people's judgment about how fatigued they are and how it affects their performance.

In one accident where fatigue was said to play a role, investigators found how "the captain and the first officer had been continuously awake for at least 16 hours. Research indicates that the normal waking day is between 14 and 16 hours and that lapses in vigilance increase and become longer if the normal waking day is extended.

In addition, the Safety Board's 1994 study of flight crew-related major aviation accidents found that flight crews that had been awake for an average of about 13 hours made significantly more errors, especially procedural and tactical decision errors, than crews that had been awake for an average of about five hours ... The "accident time was nearly two hours after the time that both pilots went to bed the night before the accident and the captain's routine bedtime ...

Research indicates that the ability to consider options decreases as people who are fatigued become fixated on a course of action ... Also, automatic processes (such as radio calls and routine behavior) are affected less by fatigue than controlled processes (such as more complex behavior, responses to new situations, and decision making). Further, fatigue deteriorates performance on work-paced tasks that are characterized by time pressure and task-dependent sequences of activities."[3]

As indicated in the example above, fatigue can be the result of a number of factors (or a combination), for example:

- Workload intensity (or, conversely, inactivity) and sustained effort (performance deteriorates as a function of "time on task");
- Physical and psychological exertion;
- Sleep deprivation or other sleep disturbances;
- Time of day effects;
- Cyrcadian desynchronization (e.g. jet lag).

While performance effects of fatigue are often difficult to prove, some of the following effects have been linked to fatigue:

- Vigilance effects. Tasks requiring sustained attention or quick reaction times are particularly vulnerable;
- Cognitive slowing. Slower responses on all kinds of cognitive tasks, such as reasoning, tracking, arithmetic, generating options and making decisions;
- Memory effects. Build-up of memory compromised by attention deficits (and possibly lapsing);
- Lapsing. Also called "microsleeps" (from ten seconds in duration), where people do not respond to external stimuli.

Fatigue is a huge research area, and one result is that most safety-critical professions are actually regulating worktime limitations. These limits, however, are often more the outcome of industrial or political consensus, than they are solidly founded on research data. And duty time limitations, while an enormous step forward in managing fatigue that worlds such as medicine could still learn from, are not always able to deal with the inevitable consequences of operating during unusual periods across timezones:

"After more than a decade of research, NASA said in a 1994 report that weariness from long shifts, irregular schedules and frequent timezone changes could be exacerbated by inactivity during long, uninterrupted flights in largely automated cockpits. 'Sleepiness can degrade essentially every aspect of human performance,' NASA researchers said.

Eighty-nine per cent of 1424 flight crew members from commuter airlines surveyed by NASA identified fatigue as a moderate or serious concern. Similar sentiments were expressed by cockpit crews of major carriers in reports gathered by NASA ...

'After seven days in the Pacific, we fly all night from Bangkok to Narita, have a short day layover, then fly all night to Honolulu', one pilot wrote. 'Some or all of the crew passes out on the last leg from fatigue. We are so tired by the approach and landing that our thinking and reaction times are similar to being drunk. If the weather wasn't consistently good in Honolulu, I'm sure we would have lost an airplane a long time ago.'"[4]

Buggy and Inert Knowledge

Practitioners usually do not come to their jobs unprepared. They possess a large amount of knowledge in order to manage their operations. The application of knowledge, or using it in context, is not a straightforward activity, however. In order to apply knowledge to manage situations, people need three things:

- People need to possess the knowledge;
- People need to have that knowledge organized in a way that makes it usable for the situation at hand;
- People need to activate the relevant knowledge in context.

While each presents unique challenges and possible difficulties, the three interrelate. You may be able to pick out features of people's knowledge (and how it was taught and applied) to help you understand human error.

So first, practitioners need to possess the knowledge. Ask yourself whether the right knowledge was there, or whether it was erroneous or incomplete. People may have been trained in ways that leave out important bits and pieces. The result is buggy knowledge, or a buggy mental model (with gaps, holes and bugs in it).

There are accounts of pilots trying to take off in snowy weather who rely on the exhaust blast of the aircraft in front of them to clear the latest snow off the runway. This may be a case of pressure to take off (see the next chapter) without waiting for plows to clear the snow, but also in part a case of buggy knowledge about how effective that clearing is. The question is not how pilots can be so silly to rely on this, but how they have been led to believe (built up a mental model) that this is actually effective.

Second, practitioners need to have the knowledge organized in a way that makes it useable for the situation at hand. People may have learned how to deal with complex systems or complex situations by reading books or manuals about them. This does not guarantee that the knowledge is organized in a way that allows them to apply it effectively in operational circumstances.

The way in which knowledge is organized in the head seriously affects people's ability to perform well. Knowledge organization is in turn a result of how the material is taught or acquired. Feltovich[5] has investigated how knowledge can be wrongly organized, especially in medicine, leading to misconceptions and misapplications.

One example is that students have learned to see highly interconnected processes as independent from one another, or to treat dynamic processes as static, or to treat multiple processes as the same thing, since that is how they were taught. For example, changes in cardiac output (the rate of blood flow, which is the change of position of volume/minute), are often seen as though they were changes in blood volume. This would lead a student to believe that increases in cardiac output could propagate increases of blood volume, and consequently blood pressure, when in fact increases in blood flow decreases pressure in the veins.

Another example where knowledge organization that can mismatch the application situation happens in Problem-Based Learning (PBL). Popular in many circles, it carries a risk that students will see one instance of a problem they are confronted with in training as canonical for all instances they will encounter subsequently. This is overgeneralization: treating different problems as similar.

Third, practitioners also need to activate the relevant knowledge, that is, bring it to bear in context. People can often be shown to possess the knowledge necessary for solving a problem (in a classroom situation, where they are dealing with a textbook problem), but that same knowledge won't "come to mind" when needed in the real world; it remains inert.

If material is learned in neat chunks and static ways (books, most computer-based training) but needs to be applied in dynamic situations that call for novel and intricate combinations of those knowledge chunks, then inert knowledge is a risk. In other words, when you suspect inert knowledge, look for mismatches between how knowledge is acquired and how it is (to be) applied.

Training practitioners to work with automation is difficult. Pilots, for example, who learn to fly automated airplanes typically learn how to work the computers, rather than how the computers actually work. They learn the input-output relationships for various well-developed and common scenarios, and will know which buttons to push when these occur on the line. Problems emerge, however, when novel or particularly difficult situations push pilots off the familiar path, when circumstances take them beyond the routine. Knowledge was perhaps once acquired and demonstrated about automation modes or configurations that are seldom used. But being confronted with this in practice means that the pilot may not know what to do—knowledge that is in principle in the head, will remain inert.

New Technology and Computerization

Human work in safety-critical domains has almost without exception become work with technology. This means that human-technology interaction is an increasingly dominant source of error. Technology has shaped and influenced the way in which people make errors. It has also affected people's opportunities to detect or recover from the errors they make and thus, in cases, accelerated their journeys towards breakdown.

As is the case with organizational sources of error, human-technology errors are not random. They too are systematically connected to features of the tools that people work with and the tasks they have to carry out. Here is a guide,[6] first to some of the "errors" you may typically find. Then a list of technology features that help produce these errors, and then a list of some of the cognitive consequences of new technology that lie behind the creation of those errors.

More can, and will, be said about technology. The section pays special attention to automation surprises, since these appear to form a common pattern of failure underneath many automation-related mishaps. We will also look at how new technology influences coordination and operational pressures in the workplace.

The New View on the role of technology

How does the New View on human error look at the role of technology? New technology does not remove the potential for human error, but changes it. New technology can give a system and its operators new capabilities, but it inevitably brings new complexities too:

- New technology can lead to an increase in operational demands by allowing the system to be driven faster; harder; longer; more precisely or minutely. Although first introduced as greater protection against failure (more precise approaches to the runway with a Head-Up-Display, for example), the new technology allows a system to be driven closer to its margins, eroding the safety advantage that was gained.
- New technology is also often ill-adapted to the way in which people do or did their work, or to the actual circumstances in which people have to carry out their work, or to other technologies that were already there.
- New technology often forces practitioners to tailor it in locally pragmatic ways, to make it work in real practice.
- New technology shifts the ways in which systems break down.
- It asks people to acquire more knowledge and skills, to remember new facts.
- It adds new vulnerabilities that did not exist before. It can open new and unprecedented doors to system breakdown.

The New View of human error maintains that:

- People are the only ones who can hold together the patchwork of technologies introduced into their worlds; the only ones who can make it all work in actual practice;
- It is never surprising to find human errors at the heart of system failure because people are at the heart of making these systems work in the first place.

Typical errors with new technology

If people were interacting with computers in the events that led up to the mishap, look for the possibility of the following "errors":

- **Mode error.** The user thought the computer was in one mode, and did the right thing had it been in that mode, yet the computer was actually in another mode.
- **Getting lost** in display architectures. Computers often have only one or a few displays, but a potentially unlimited number of things you can see on them. Thus it may be difficult to find the right page or data set.
- **Not coordinating computer entries.** Where people work together on one (automated) process, they have to invest in common ground by telling one another what they tell the computer, and double-checking each other's

work. Under the pressure of circumstances or apparently meaningless repetition, such coordination may not happen consistently.
- **Workload**. Computers are supposed to off-load people in their work. But often the demand to interact with computers concentrates itself on exactly those times when there is already a lot to do; when other tasks or people are also competing for the operator's attention. You may find that people were very busy programming computers when other things were equally deserving of their attention.
- **Data overload**. People may be forced to sort through large amounts of data produced by their computers, and may be unable to locate the pieces that would have revealed the true nature of their situation. Computers may also spawn all manner of automated (visual and auditory) warnings which clutter a workspace and proliferate distractions.
- **Not noticing changes**. Despite the enormous visualization opportunities the computer offers, many displays still rely on raw digital values (for showing rates, quantities, modes, ratios, ranges and so forth). It is very difficult to observe changes, trends, events or activities in the underlying process through one digital value clicking up or down. You have to look at it often or continuously, and interpolate and infer what is going on. This requires a lot of cognitive work by the human.
- **Automation surprises** are often the end result: the system did something that the user had not expected. Especially in high tempo, high workload scenarios, where modes change without direct user commands and computer activities are hard to observe, people may be surprised by what the automation did or did not do (see next section).

Computer features

What are some of the features of today's technology that contribute systematically to the kinds of errors discussed above?

- Computers can make things "invisible"; they can hide interesting changes and events, or system anomalies. The presentation of digital values for critical process parameters contributes to this "invisibility". The practice of showing only system *status* (what mode it is in) instead of *behavior* (what the system is actually doing; where it is going) is another reason. The interfaces may look simple or appealing, but they can hide a lot of complexity.
- Computers, because they only have one or a few interfaces (this is called the "keyhole problem"), can force people to dig through a series of display

pages to look for, and integrate, data that really are required for the task in parallel. A lot of displays is not the answer to this problem of course, because then navigation across displays becomes an issue. Rather, each computer page should present aids for navigation (How did I get here? How do I get back? What is the related page and how do I get there?). If not, input or retrieval sequences may seem arbitrary, and people will get lost.

- Computers can force people into managing the interface (How do I get to that page? How do we get it into this mode?) instead of managing the safety-critical process (something the computer was promised to help them do). These extra interface management burdens often occur during periods of high workload.
- Computers can change mode autonomously or in other ways that are not commanded by the user (these mode changes can for example result from pre-programmed logic, much earlier inputs, inputs from other people or parts of the system, and so forth).
- Computers ask people typically in the most rudimentary or syntactic ways to verify their entries (Are you sure you want to go to X? We'll go to X then) without addressing the meaning of their request and whether it makes sense given the situation. And when people tell computers to proceed, it may be difficult to make them stop. All this limits people's ability to detect and recover from their errors.
- Computers are smart, but not that smart. Computers and automation can do a lot for people—they can almost autonomously run a safety-critical process. Yet computers may know little about the larger situation around them. Computers sometimes assume a stable world where they can proceed with their pre-programmed routines even if these are inappropriate. They can dutifully execute user commands that make no sense given the situation; they can interrupt people's other activities without knowing they are bothersome.

Cognitive consequences of computerization

The characteristics of computer technology discussed above shape the way in which people assess, think, decide, act and coordinate, which in turn determines the reasons for their "errors":

- Computers increase demands on people's memory (What was this mode again? How do we get to that page?).
- Computers ask people to add to their package of skills and knowledge for managing their processes (How to program, how to monitor, and so forth).

So What Went Wrong? 151

Training may prove no match to these new skill and knowledge requirements: much of the knowledge gained in formal training may remain inert (in the head, not practically available) when operators get confronted with the kinds of complex situations that call for its application.

- Computers can complicate situation assessment (they may show digital values or letter codes instead of system behavior) and undermine people's attention management (how you know where to look when).
- By new ways of representing data, computers can disrupt people's traditionally efficient and robust scanning patterns.
- Through the limited visibility of changes and events, the clutter of alarms and indications, extra interface management tasks and new memory burdens, computers increase the risk of people falling behind in high tempo operations.
- Computers can increase system reliability to a point where mechanical failures are rare (as compared with older technologies). This gives people little opportunity for practicing and maintaining the skills for which they are, after all, partly still there: managing system anomalies.
- Computers can undermine people's formation of accurate mental models of how the system and underlying process work, because working the safety-critical process through computers only exposes them to a superficial and limited array of experiences.
- Computers can mislead people into thinking that they know more about the system than they really do, precisely because the full functionality is hardly ever shown to them (either in training or in practice). This is called the knowledge calibration problem.
- Computers can force people to think up strategies (programming "tricks") that are necessary to get the task done. These tricks may work well in common circumstances, but can introduce new vulnerabilities and openings to system breakdown in others.

New technology and operational pressures

The introduction of new technology can increase the operational requirements and expectations that organizations impose on people. Organizations that invest in new technologies often unknowingly exploit the advances by requiring operational personnel to do more, do it more quickly, do it in more complex ways, do it with fewer other resources, or do it under less favorable conditions.

Larry Hirschorn talks about a law of systems development, which is that every system always operates at its capacity. Improvements in the form of new technology get stretched in some way, pushing operators back to the edge of the operational envelope from which the technological innovation was supposed to buffer them.

In operation Desert Storm, during the first Gulf War, much of the equipment employed was designed to ease the burden on the operator, reduce fatigue, and simplify the tasks involved in combat. Instead these advances were used to demand more from the operator. Not only that, almost without exception, technology did not meet the goal of unencumbering the military personnel operating the equipment. Weapon and support systems often required exceptional human expertise, commitment and endurance. The Gulf War showed that there is a natural synergy between tactics, technology and human factors: effective leaders will exploit every new advance to the limit.[7]

Automation Surprises

Automation surprises are cases where people thought they told the automation to do one thing, while it is actually doing another. For example, the user dials in a flight path angle to make the aircraft reach the runway, whereas the automation is actually in vertical speed mode—interpreting the instruction as a much steeper rate of descent command rather than a flight angle. Automation may be doing something else because of many reasons, among them:

- It is in a different mode from what people expected or assumed when they provided their instructions.
- It shifted to another mode after having received instructions.
- Another human operator has overriden the instructions given earlier.

Automation surprises appear to occur primarily when the following circumstances are present:

- Automated systems act on their own, that is, without immediately preceding user input. The input may have been provided by someone else, or a while back, or the change may be the result of pre-programmed system logic.
- There is little feedback about the behavior of the automated system that would help the user discover the discrepancy. Feedback is mostly status-based, telling the user—in cryptic abbreviations—what state the system is in ("I am now in V/S"), not what it is actually doing or what it will do in the near future. So even though these systems behave over time, they do not tell the user of their behavior, only about their status.

- Event-driven circumstances (where the unfolding situation governs how fast people need to think, decide, act) often help create automation surprises. Novel situations, ones that people have not encountered before, are also more likely to trigger automation surprises.
- Intervening in the behavior after an automation surprise may also be hard. In addition to being "silent" (not very clear about their behavior), automation is often hard to direct: it is unclear what must be typed or tweaked to make it do what the user wants. People's attention shifts from managing the process to managing the automation interface.

People can have a hard time discovering that automation is not behaving according to their intentions. There is little evidence that people are able to pick the mismatch up from the displays or indications that are available—again because they are often status-oriented and tiny. People will discover that the automation has been doing something different when they first notice strange or unexpected process behavior. Serious consequences may have already ensued by that time.

Automation and coordination

In efforts to sum up the issues with automation, we often refer to people slipping "out-of-the-loop". We think, in our folksy ways, that people performing duties as system monitors will be lulled into complacency, lose situational awareness, and not be prepared to react in a timely manner when the system fails.

There is, however, little evidence that the out-of-the-loop problem leads to the serious accidents. First, automation hardly ever fails in a binary sense. In fact, manufacturers consistently point out in the wake of accidents how their automation behaved as designed. An out-of-the-loop problem—in the sense that people are unable to intervene effectively "when the system fails" after a long period of only monitoring—does not lie behind the problems that occur. In fact, the opposite appears true.

Bainbridge wrote about the ironies of automation in 1987.[8] She observed that automation took away the easy parts of a job, and made the difficult parts more difficult:

- Automation relies on human monitoring, but people are bad at monitoring for very infrequent events.
- Indeed, automation does not fail often, which limits people's ability to practice the kinds of breakdown scenarios that still justifies their marginal presence in the system.

Here, the human is painted as a passive monitor, whose greatest safety risks would lie in deskilling, complacency, vigilance decrements and the inability to intervene in deteriorating circumstances.

These problems occur, obviously, but these are not the behaviors that precede accidents with, for example, automated airliners or other forms of process control over the past two decades. Instead, people have roles as active supervisors, or managers, who need to coordinate a suite of human and automated resources in order to get their process to work. Yet people's ability to coordinate their activities with those of computers and other people is made difficult by silent and strong (or powerful and independent) automation; by the fact that people may have private access to the automation (e.g. each pilot has his/her own flight management system control/display unit); and because demands to coordinate with the automation accrue during busy times when a lot of communication and coordination with other human operators is also needed.

The same features, however, that make coordination difficult make it critically necessary. This is where the irony lies. Coordination is necessary to invest in a shared understanding of what the automated system has been told to do (yet difficult because it can be told to do things separately by any pilot and then go on its way without showing much of what it is doing). Coordination is also necessary to distribute work during busier, higher pressure operational episodes, but such delegation is difficult because automation is hard to direct and can shift a pilot's attention from flying the aircraft to managing the interface.

Accidents are preceded by practitioners being active managers—typing, searching, programming, planning, responding, communicating, questioning—trying to coordinate their intentions and activities with those of other people and the automation, exactly like they would in the pursuit of success and safety. What seems to lie behind these mishaps is a breakdown in coordination, or teamplay, between humans and automated systems that can eventually turn a manageable situation into an unrecoverable one.

Procedural Adaptations

Why don't people follow the rules? Systems would be so much safer if they did. This is often our naive belief. Procedure-following equals safety. But reality is not quite so simple.

Here is an interesting thought experiment. A landing checklist has all the things on it that need to be done to get an aircraft ready for landing. Pilots read the items off it, and then make them happen in the systems around them. For example:

- *Hydraulic pumps*   ON
- *Altimeters*   SET
- *Flight Instruments* SET
- *Seat Belt sign*   ON
- *Tail de-ice*   AS REQ.
- *Gear*   DOWN
- *Spoilers*   ARMED
- *Auto brakes*   SET
- *Flaps and slats*   SET

There is no technical reason why computers could not accomplish these items today. Automating a before-landing checklist is, as far as software programming goes, entirely feasible. Why do we rely on unreliable people to accomplish these items for us? The answer is context. Not every approach is the same. The world is complex and dynamic. Which items come when, or which ones can or must be accomplished, perhaps ahead of others, is something that is determined in part by the situation in which people find themselves.

Procedures are not sufficiently sensitive to the many subtle variations in context, which is why people need to interpret them. Going through that list rigidly and uncompromisingly (precisely the way you may think people should have acted when you sort through the rubble of their mishap), could lead to people ignoring features of the situation that make accomplishing a particular item at that time entirely inappropriate.

Applying procedures is not simple rule-following. Applying procedures successfully in actual situations is a substantive cognitive activity.

Think for example of an inflight fire or other serious malfunction where pilots must negotiate between landing overweight or dumping fuel (two things you simply can't do at the same time), while sorting through procedures that aim to locate the source of trouble—in other words, doing what the book and training and professional discipline tells them to do. If the fire or malfunction catches up with the pilots while they are still airborne, you may say that they should have landed instead of bothered with anything else. But it is only hindsight that allows you to say that.

Situations may (and often do) occur where multiple procedures need to be applied at once, because multiple things are happening at once. But items in these various procedures may contradict one another. There was one case, for example, where a flight crew noticed both smoke and vibrations in their cockpit. There was no procedure that told them how to deal with the combination of symptoms. Adaptation, improvisation was necessary to deal with the situation.

There may also be situations, like the one described above, where hindsight tells you that the crew should have adapted the procedure, shortcut it, abandoned it. But they failed to adapt; they stuck to the rules rigidly. So the fire caught up with them. Then there are standard situations where rigid adherence to procedures leads to less safety. Applying procedures can thus be a delicate balance between:

- Adapting procedures in the face of either unanticipated complexity or a vulnerability to making other errors. But people may not be entirely successful at adapting the procedure, at least not all of the time. They will then be blamed for not following procedures; for improvising.
- Sticking to procedures rigidly, and discovering that adapting them would perhaps have been better. People will then be blamed for not being flexible in the face of a need to be so.

Notice the double bind that practitioners are in here: whether they adapt procedures or stick with them, with hindsight they can get blamed for not doing the other thing. Organizations often react counterproductively to the discovery of either form of trouble:

- When procedures are not followed but should have been, they may send more exhortations to follow procedures (or even more procedures) into the operation.
- When procedures were applied rigidly and followed by trouble, organizations may send signals to give operational people more discretion.

But none of this resolves the fundamental double bind people are in, in fact, it may even tighten the bind. Either signal only shifts people's decision criterion a bit—faced with a difficult situation, the evidence in people's circumstances should determine whether they stick to the rules or adapt. But where the decision lies, depends in large part on the signals the organization has been sending.

Two opposite models of procedures

So there really are two different ways in which you can think about procedures. The first one is rather OldView, and not really sensitive to how people actually accomplish safety-critical work in practice. The second one acknowledges how procedures are but one resource for action (among many others).

Table 14.2 Opposing models of procedures and safety.

Model 1 (Old View)	Model 2 (New View)
Procedures are the best thought-out, safest way to carry out a task	Procedures are resources for action (next to other resources)
Procedure-following is IF-THEN, rule-based behavior	Applying procedures successfully is a substantive, skillful cognitive activity
Safety results from people following procedures	Procedures cannot guarantee safety. Safety comes from people being skillful at judging when and how they apply
Safety improvements come from organizations telling people to follow procedures and enforcing this	Safety improvements come from organizations monitoring and understanding the gap between procedures and practice

Remember the error trap of not arming the ground spoilers on an aircraft before landing. This shows that people actually create safety by adapting procedures in a locally pragmatic way. The pilot who adapted successfully was the one who, after years of experience on this particular aircraft type, figured out that he could safely arm the spoilers four seconds after "gear down" was selected, since the critical time for potential gear compression was over by then. He had refined a practice whereby his hand would go from the gear lever to the spoiler handle slowly enough to cover four seconds—but it would always travel there first. He then had bought himself enough time to devote to subsequent tasks such as selecting landing flaps and capturing the glide slope. This obviously "violates" the original procedure, but the "violation" is actually an investment in safety, the creation of a strategy that help forestall failure.

"Procedure violation", or "non-compliance" are obviously counterproductive and judgmental labels—a form of saying "human error" all over again, without explaining anything. These labels simply rehearse that people should stick with the rules, and then everything will be safe. Ever heard of the "work to rule strike"? This is when people, instead of choosing to stop work altogether, mobilize industrial action by following all the rules for a change. What typically happens? The system comes to a gridlock. Follow all the rules by the book, and your system will no longer work.

Labels such as violations miss the complexity beneath the successful application, and adaptation, of procedures, and may lead to misguided countermeasures. High reliability organizations do not try to constantly close the gap between procedures and practice by exhorting people to stick to the rules. Instead, they continually invest in their understanding of the reasons beneath the gap. This is where they try to learn—learn about ineffective guidance; learn about novel, adaptive strategies and where they do and do not work. The next chapter will say more about this.

Notes

1 Neisser, U. (1976). *Cognition and reality*. San Francisco: W.H. Freeman,.
2 Orasanu, J., Martin, L., and Davison, J. (in press). Cognitive and contextual factors in aviation accidents: Decision errors. In E. Salas and G. Klein (eds). *Applications of naturalistic decision making*. Mahwah, NJ: Lawrence Erlbaum Associates.
3 National Transportation Safety Board (2001). *Aircraft accident report: Runway overrun during landing, American Airlines flight 1420, MD-82, Little Rock, Arkansas, June 1 1999 (AAR-01/02)*. Washington, DC: NTSB.
4 International Herald Tribune, 12 October 2000.
5 Feltovich, P.J., Spiro, R.J., and Coulson, R.L. (1993). Learning, teaching, and testing for complex conceptual understanding. In N. Fredericksen, R. Mislevy, and I. Bejar (eds). *Test theory for a new generation of tests*. Hillsdale, NJ: Lawrence Erlbaum Associates.
6 Material for this section comes from Woods, D.D., Johanssen, L.J., Cook, R.I., and Sarter, N.B. (1994). *Behind human error: Cognitive systems, computers and hindsight*. Dayton, OH: CSERIAC, and Dekker, S.W.A., and Hollnagel, E. (eds) (1999). *Coping with computers in the cockpit*. Aldershot, UK: Ashgate Publishing.
7 Cordesman, A.H., and Wagner, A.R. (1996). *The lessons of modern war, Vol. 4: The Gulf war*. Boulder, CO: Westview Press.
8 Bainbridge, L. (1987). Ironies of automation. In J. Rasmussen, K. Duncan, and J. Leplat (eds). *New technology and human error*. Chichester, UK: Wiley.

15 Look into the Organization

A human error problem is an organizational problem. Not because it creates problems for the organization. It is organizational because a human error problem is created, in large part, by the organization in which people work. This means that understanding human error hinges on understanding the organizational context in which people work. There are many ways to parse this context, but this chapter looks at the following:

- Procedural drift and different images of work;
- Production pressure and goal conflicts;
- Safety culture.

Procedural Drift

There are two ways of looking at a mismatch between procedures and practice:

- As non-compliant behavior, or "violations". People who violate the procedures put themselves above the law. Violations are a sign of deviance. "Violation" is a label that you impose from the outside. From that position you can see a mismatch between what people do locally and what they are supposed to do, according to your understanding of rules governing that portion of work.
- As compliant behavior. Even though actual performance may mismatch written guidance, people's behavior is typically in compliance with a complex of norms, both written and implicit. Getting the job done in time or with fewer resources may be a mark of expertise against which real performance is scored, both by peers and superiors. This may be more important (even to management) than sticking with all applicable written rules. In this case you want to understand, from the inside, how people see their behavior as conformant, not deviant.

While a mismatch between procedures and practice almost always exists, it is never constant. The mismatch can grow over time, increasing the gap between how the system was designed (or imagined) and how it actually works. This is called practical drift: the slow, incremental departure from initial written guidance on how to operate a system. This is what lies behind practical drift:

- Rules that are overdesigned (written for tightly coupled situations, for the worst case) do not match actual work most of the time. In real work, there is slack: time to recover, opportunity to reschedule and get the job done better or more smartly. This mismatch creates an inherently unstable situation that lays the basis for drift.
- Emphasis on local efficiency or cost effectiveness pushes operational people to achieve or prioritize one goal or a limited set of goals (e.g., customer service, punctuality, capacity utilization). Such goals are typically easily measurable (e.g., customer satisfaction, on-time performance), whereas it is much more difficult to measure how much is borrowed from safety.
- Past success is taken as guarantee of future safety. Each operational success achieved at incremental distances from the formal, original rules can establish a new norm. From here a subsequent departure is once again only a small incremental step. From the outside, such fine-tuning constitutes incremental experimentation in uncontrolled settings. On the inside, incremental nonconformity is an adaptive response to scarce resources, multiple goals, and often competition.
- Departures from the routine become routine. Seen from the inside of people's own work, violations become compliant behavior. They are compliant with the emerging, local ways to accommodate multiple goals important to the organization (maximizing capacity utilization but doing so safely; meeting technical requirements, but also deadlines).

Breaking the rules to get more recruits
Some say cheating needed to fill ranks

New York Times, May 4, 2005

It was late September when the 21-year-old man, fresh from a psychiatric ward, showed up at a US Army recruiting station. The two recruiters there quickly signed him up. Another recruiter said the incident hardly surprised him. He has been bending or breaking enlistment rules for months, he said, hiding police records and medical histories of potential recruits. His commanders have encouraged such deception, he said, because they know there is no other way to meet the Army's recruitment quotas.

'The problem is that no one wants to join', the recruiter said. 'We have to play fast and loose with the rules just to get by.' Others spoke of concealing mental health histories and police records. They described falsified documents, wallet-size cheat sheets slipped to applicants before the military's aptitude test, and commanding officers who look the other way. And they voiced doubts about the quality of troops destined for combat duty.

Two years ago, a policy was ended that nearly always dismissed serious offenders from recruiting. The Army's commander of recruiting explained how 'his shift in thinking was that if an individual was accused of doctoring a high-school diploma, it used to be an open-and-shut case. But now he looks at the person's value to the command first.'

Recruiting has always been a difficult job, but the temptation to cut corners is particularly strong today, as deployments in Iraq and Afghanistan have created a desperate need for new soldiers, and as the Army has fallen short of its recruitment goals in recent months. Says one expert: 'The more pressure you put on recruiters, the more likely you'll be to find people seeking ways to beat the system'. Over the past months, the Army has relaxed its requirements on age and education, a move that may have led recruiters to go easier on applicants.

Figure 15.1 shows what really may be going on and why labels such as complacency or negligence or rulebreaking not only are swift judgments, but also incomplete. They hide the really interesting dynamic, the historical perspective in which the behavior may make sense (and will continue to make sense to people on the inside, independent of what ugly words you call it).

Departures from some norm or routine may at any one moment seem to occur because people are not motivated to do otherwise. Or because people have become reckless and focused only on production goals. But departures from the routine can become the routine as a result of a much more complex picture that blends organizational preferences (and how they are communicated), earlier success, peer pressure, and compliance with implicit expectations.

So when you discover a gap between procedures and practice:

- Recognize that it is often compliance that explains people's behavior: compliance with norms that evolved over time—not deviance. What people were doing was reasonable in the eyes of those on the inside of the situation,

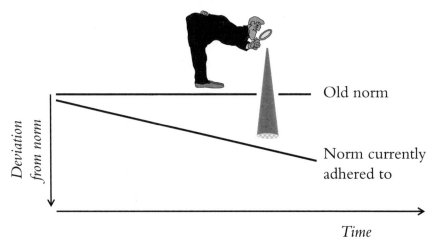

Figure 15.1 At a particular moment in time, behavior that does not live up to some standard may look like complacency or negligence. But deviance may have become the new norm across an entire operation or organization.

given the pressures and priorities operating on them and others doing the same work every day.
- Find out what organizational history or pressures exist behind these routine departures from the routine; what other goals help shape the new norms for what is acceptable risk and behavior.
- Understand that the rewards of departures from the routine are probably immediate and tangible: happy customers, happy bosses or commanders, money made, and so forth. The potential risks (how much did people borrow from safety to achieve those goals?) are unclear, unquantifiable or even unknown.
- Realize that continued absence of adverse consequences may confirm people in their beliefs (in their eyes justified!) that their behavior was safe, while also achieving other important system goals.

The incremental nature of drift is really important. While the end result of drift may seem a very large departure from originally accepted routines, each step along the way is typically only a small increment (or decrement) from what was accepted previously. As such, the small departure from the previous routine does not normally concern people that much, especially if adverse consequences do not occur either.

H.C. Andersen tells a wonderful fairytale; a story of drift avant la lettre, a children's story of an incremental slide into bad judgment. Each step of the way seemed to make perfect sense in and of itself, and only a small deviation—if at all—from what had previously been accepted. It was only the aggregate picture that became increasingly absurd—seeming to make no sense at all.

A farmer was asked by his wife to trade their only horse for something they could have more use of. The man rode to town but came upon another farmer with a cow. 'Let's talk', the man said. 'I want to trade my horse for your cow. A cow makes much more sense than a horse'. Milk, labor, meat—perfect. The man glowed with pride and wanted to go home with his new cow. But hang on, he was on his way to the market anyway. Might as well proceed. There he ends up trading the cow for a sheep (wool to keep warm), the sheep for a goose (for garbage disposal), the goose for a chicken (eggs), and the chicken for a bag of rotten apples (because his own apple tree delivered a poor harvest last time).

The trade from a horse down to some rotten apples is a very bad trade indeed. It makes no sense (fortunately, H.C. Andersen gave this man a very understanding and forgiving wife). But this wasn't the trade at all. The man made a series of small trades, each one sensible and reasonable when held up against local criteria at the time. It is only the aggregate that makes fairytale one of a colossal slide into bad judgment. Most ingredients of this slide are present in the drift into failure in organizational life as well.

Borrowing from safety

Deviations from the norm can themselves become the norm

With rewards constant and tangible, departures from the routine may become routine across an entire operation or organization. Without realizing it, people start to borrow from safety, and achieve other system goals because of it—production, economics, customer service, political satisfaction. Behavior shifts over time because other parts of the system send messages, in subtle ways or not, about the importance of these goals. In fact, organizations reward or punish operational people in daily trade-offs ("We are an ON-TIME operation!"), focusing them on goals other than safety. The lack of adverse consequences with each trade-off that bends to goals other than safety, strengthens people's tacit belief that it is safe to borrow from safety.

In The Challenger Launch Decision, *Diane Vaughan has carefully documented how an entire organization started borrowing from safety—reinforced by one successful Space Shuttle Launch after the other, even if O-rings in the solid rocket boosters showed signs of heat damage. The evidence for this O-ring "blow-by" was each time looked at critically, assessed against known criteria, and then decided upon as acceptable. Vaughan has called this repeated process "the normalization of deviance": what was deviant earlier, now became the new norm. This was thought to be safe: after all, there were two O-rings: the system was redundant. And if past launches were anything to go by (the most tangible evidence for success), future safety would be guaranteed or at least highly likely. The Challenger Space Shuttle, launched in cold temperatures in January 1986, showed just how much NASA had been borrowing from safety: it broke up and exploded after lift-off because of O-ring blow-by.*[1]

The problem with complex, dynamic worlds is that safety is not a constant. Past success while departing from a routine is not a guarantee for future safety. In other words, a safe outcome today is not a guarantee of a safe outcome tomorrow, even if behavior is the same. Circumstances change, and so do the safety treats associated with them. Doing what you do today (which could go wrong but did not) does not mean you will get away with it tomorrow. The dynamic safety threat is pictured in the figure overleaf.

Murphy's law is wrong. What can go wrong usually goes right. But then we draw the wrong conclusion

While people inside organizations adjust their behavior to accommodate other system pressures (e.g. on-time performance), safety threats vary underneath, setting them up for problems sometime down the line. Of course, in order to be wrong, Murphy's law has to be right—in the end, that is. That which can go wrong must at some point go wrong, otherwise there would not even be a safety threat.

Production Pressure and Goal Conflicts

A major engine behind routine divergence from written guidance is the need to pursue multiple goals simultaneously. Multiple goals mean goal conflicts. In

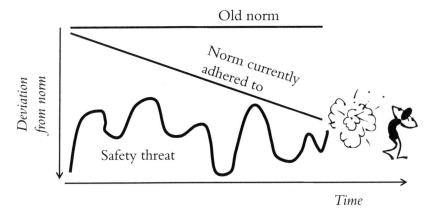

Figure 15.2 Murphy's law is wrong.[2] What can go wrong usually goes right, and over time organizations can come to think that a safety threat does not exist or is not so bad.

most work, contradictory goals are the rule, not the exception. Any human factors investigation that does not take goal conflicts seriously, does not take human work seriously.

A study of flight dispatchers illustrated a basic dilemma. Would bad weather hit a major hub airport or not? What should the dispatchers do with all the airplanes en route? Safety (by making aircraft divert widely around the weather) would be a pursuit that "tolerates a false alarm but deplores a miss". In other words, if safety is the major goal, then making all the airplanes divert even if the weather would not end up at the hub (a false alarm) is much better than not making them divert and sending them headlong into bad weather (a miss). Efficiency, on the other hand, severely discourages the false alarm, whereas it can actually deal with a miss.

This is the essence of most operational systems. Though safety is a (stated) priority, these systems do not exist to be safe. They exist to provide a service or product, to achieve economic gain, to maximize capacity utilization. But still they have to be safe. One starting point, then, for understanding a driver behind routine deviations, is to look deeper into these goal interactions, these basic incompatibilities in what people need to strive for in their work. If you want to understand human error, you need to find out how people themselves view these conflicts from inside their operational reality, and how this contrasts with other views of the same activities (e.g. management, regulator, public).

NASA's "Faster, Better, Cheaper" organizational philosophy in the late 1990s epitomized how multiple, contradictory goals are simultaneously present and active in complex systems. The loss of the Mars Climate Orbiter and the Mars Polar Lander in 1999 were ascribed in large part to the irreconcilability of the three goals (faster and better and cheaper), which drove down the cost of launches, made for shorter, aggressive mission schedules, eroded personnel skills and peer interaction, limited time, reduced the workforce, and lowered the level of checks and balances normally found. People argued that NASA should pick any two from the three goals. Faster and cheaper would not mean better. Better and cheaper would mean slower. Faster and better would be more expensive. Such reduction, however, obscures the actual reality facing operational personnel in safety-critical settings. These people are there to pursue all three goals simultaneously—to make it faster, better, and cheaper.

Some organizations pass on their goal conflicts to individual practitioners quite openly, and this makes your understanding of human error rather straightforward. Some airlines, for example, pay their crews a bonus for on-time performance.

An aviation publication commented on one of those operators (a new airline called Excel, flying from England to holiday destinations): "As part of its punctuality drive, Excel has introduced a bonus scheme to give employees a bonus should they reach the agreed target for the year. The aim of this is to focus everyone's attention on keeping the aircraft on schedule."[3]

Such plain acknowledgement of goal priorities, however, is not common. Most important goal conflicts are never made so explicit, arising rather from multiple irreconcilable directives from different levels and sources, from subtle and tacit pressures, from management or customer reactions to particular trade-offs.

For example, the operating manual of a large international airline opens by stating that "(1) our flights shall be safe; (2) our flights shall be punctual; (3) our customers will find value for money." In other words, the flights have to be better, faster and cheaper. All at the same time.

The actual managing of goal conflicts under uncertainty gets pushed down into local operating units—control rooms, cockpits, operating theaters and the like. There the conflicts are to be negotiated and resolved in the form of thousands of little and larger daily decisions and trade-offs. These are no longer decisions and trade-offs made by the organization, but by individual operators or crews.

The gap between work-as-imagined and work-as-done

Different images of work can exist side-by-side in an organization. One is the formal image, of work-as-imagined. The other is the actual image, about work as it is done in reality. These images may diverge substantially. Look at Table 15.1. This table displays the two different images of disaster relief work. Should relief workers in the field follow procedure and protocol and listen to their head offices? Or should they concentrate on getting the job done locally? This may involve significant improvisation.

Table 15.1 Two images of disaster relief work

Formal image of relief work	Actual image of relief work
Allegiance to distant supervisors	Dissociation from distant supervisors
Higher-order goals overriding in local decisions	Higher order goals less critical
Plans overspecified	Plans brittle in face of contingencies
Adherence to procedure and protocol	Drift from procedure and protocol as soon as hit field of practice
Deference to hierarchy and organizational structure	Deference to those with local resources and experience
Constrained by national and organizational boundaries	Improvisation across national and team boundaries
Authority lines clear (on paper)	Authority lines diffuse in reality
Ask before acting	Tell after acting (maybe)

One problem with relief work is its political dimension: what may solve a problem locally can create an embarrassment globally (as another country was supposed to get there first, or the disaster country cannot be seen to accept help from a particular donor-country). So relief workers need to accept rules and bureaucratic accountability on the one hand, but need to act in an environment full of surprise on the other. This tension creates pressure for change: as soon as relief workers hit their field of practice, their real work begins to drift from the official image of it.

The problem is not that different images of work exist. Problems arise when the organization is not sufficiently aware of the gap between these images. Having a gap is not an indication of a dysfunctional organization. But not knowing about it, and not learning why it exists, is. The existence of the gap, after all, says much about where the organization thinks its sources of resilience and safety and risk lie, versus where they actually lie. The more ignorance about this gap, the more difficult it is to make effective organizational investments in safety, as you may be investing in the wrong thing.

Follow the Goal Conflicts

Practical strategies that have evolved to deal with goal conflicts form an important ingredient in this gap between how people think the system works ("safety comes first!") and how it actually works. But how do you go about following the goal conflicts in your understanding of human error? The first thing to realize is that safety is rarely the only goal that governs what people do. Most complex work is characterized by multiple goals, all of which are active or must be pursued at the same time. Depending on the circumstances, most of these goals will somehow be at odds with one another most of the time, producing goal conflicts. As said, any understanding of human error has to take the potential for goal conflicts into account.

A woman was hospitalized with severe complications of an abdominal infection. A few days earlier, she had seen a physician with complaints of aches, but was sent home with the message to come back in eight days for an ultrasound scan if the problem persisted. In the meantime, her appendix burst, causing infection and requiring major surgery. The woman's physician had been under pressure from her managed care organization, with financial incentives and disincentives, to control the costs of care and avoid unnecessary procedures.[4] *The problem is that a physician might not know that a procedure is unnecessary before doing it, or at least doing part of it. Pre-operative evidence may be too ambiguous. Physicians end up in difficult double binds, created by the various organizational pressures.*

You will quickly see that "safety" is almost always cited as an organization's overriding goal. But it is never the only goal (and in practice not even a measurably overriding goal), or the organization would have no reason to exist. People who work in these systems have to pursue multiple goals at the same time, which often results in goal conflicts. Goal trade-offs can be generated:

- By the nature of operational work itself;
- By the nature of safety and different threats to it;
- At the organizational level.

Anesthesiology presents interesting inherent goal conflicts. On the one hand, anesthesiologists want to protect patient safety and avoid being sued for malpractice afterward. This maximizes their need for patient information and pre-operative workup. But hospitals continually have to reduce costs and increase patient turnover, which produces pressure to admit, operate and discharge patients on the same day. Other pressures stem from the need to maintain smooth relationships and working practices with other professionals (surgeons, for example), whose schedules interlock with those of the anesthesiologists.[5]

The complexity of these systems, and of the technology they employ, can also mean that one kind of safety needs to be considered against another. Here is an example of a goal trade-off that results from the nature of safety in different contexts:

The Space Shuttle Challenger broke up and exploded shortly after lift-off in 1986 because hot gases bypassed O-rings in the booster rockets. The failure has often been blamed on the decision that the booster rockets should be segmented (which created the need for O-rings) rather than seamless "tubes". Segmented rockets were cheaper to produce—an important incentive for an increasingly cash-strapped operation.

The apparent trade-off between cost and safety hides a more complex reality where one kind of safety had to be traded off against another—on the basis of uncertain evidence and unproven technology. The seamless design, for example, could probably not withstand predicted prelaunch bending moments, or the repeated impact of water (which is where the rocket boosters would end up after being jettisoned from a climbing shuttle). Furthermore, the rockets would have to be transported (probably over land) from manufacturer to launch site: individual segments posed significantly less risk along the way than a monolithic structure filled with rocket fuel.[6]

Goal conflicts can be generated by the organizational or social context in which people work. The trade-off between safety and schedule is often mentioned as an example. But other factors produce competition between different goals too, for example:

- Management policies;

- Earlier reactions to failure (how has the organization responded to similar situations before?);
- Subtle coercions (to do what the boss wants, not what he or she says);
- Legal liability;
- Regulatory guidelines;
- Economic considerations (fuel usage, customer satisfaction, public image, and so forth).

Operators can also bring personal or professional interests with them (career advancement, avoiding conflicts with other groups), that enter into their negotiations among different goals.

How do you find out about goal conflicts in your investigation? Not all goals are written down in guidance or procedures or job descriptions. In fact, most are probably not. This makes it difficult to trace or prove their contribution to particular assessments or actions. To evaluate the assessments and actions of the people you are investigating, you should:

- List the goals relevant to their behavior at the time;
- Find out how these goals interact or conflict;
- Investigate the factors that influenced people's criterion setting (i.e. what and where was the criterion to pursue the one goal rather than the other, and why was it there?).

Remember the pilot from Chapter 2 who refused to fly in severe icing conditions and was subsequently fired? If you were to list the goals relevant in his decision making, you would find schedule, passenger connections, comfort, airline disruptions and, very importantly: the safety of his aircraft and passengers. In his situation, these goals obviously conflicted. The criterion setting in resolving the goal conflict (by which he decided not to fly) was very likely influenced by the recent crash of a similar aircraft in his airline because of icing.

The firing of this pilot sent a message to those who come after him and face the same trade-off. They may decide to fly anyway because of the costs and wrath incurred from chief pilots and schedulers (and passengers even). Yet the lawsuit sent another message, that may once again shift the criterion back a bit toward not flying in such a situation.

It is hard for organizations, especially in highly regulated industries, to admit that these kinds of tricky goal trade-offs arise; even arise frequently. But denying the existence of goal conflicts does not make them disappear. For a human error investigation it is critical to get these goals, and the conflicts they produce, out in the open. If not, organizations easily produce something that looks like a solution to a particular incident, but that in fact makes certain goal conflicts worse.

Safety Culture

Organizations with a "strong safety culture" could be more resilient against drifting into failure. But what is a safety culture, and what makes it strong? Much has been written about this, but little has been grasped. Common ingredients, however, seem to include:[7]

- **Management commitment**. Managers cannot expect employees to find safety more important than they do (or appear to do). If safety is low on a management priority list, this will have its effect on how it is treated in the rest of the organization.
- **Management involvement**. Managers have a real understanding of the sources of operational safety and risk, without resorting to distant idealizations of how real work takes place. They can show this involvement by participating in training or oversight.
- **Employee empowerment**. Employees feel that they can make a difference by influencing operational policies, they take pride in the safety record and feel in part responsible for creating it.
- **Incentive structures**. While incentive structures often exist for economic gain, some organizations also connect safety behavior to incentive structures (e.g. promotions).
- **Reporting systems**. An effective flow of safety-related information is the lifeblood of a safety culture. Reporting is only the beginning of course: in order to learn and improve, organizations must actually do something with the information received (which in turn hinges on the preceding points).

People sometimes make a distinction between safety culture and safety climate. Safety climate refers to a more short-term, changeable atmosphere of openness and learning that can result from immediate management action, or a crisis such as a recent incident. A culture is more enduring, representing features that emerge from the organization over a longer time. A safety culture is a culture that allows the boss to hear bad news. This presents two problems, a relatively easy one and a really hard one. People need to feel relevant and empowered to get news to

A safety culture is a culture that allows the boss to hear bad news

the boss, structures for making the news flow need to exist, and management needs to show commitment to such news and doing something about it. This is hard, of course, but not as hard as the really hard problem.

The hard problem is to decide what is bad news. Previous parts of this chapter show that an entire operation or organization can shift its idea of what is normative, and thus shift what counts as bad news. On-time performance can be normative, for example, even if it means that operators unknowingly borrow from safety to achieve it. In such cases, the hurried nature of a departure or arrival is not bad news that is worth reporting (or worth listening to, for that matter). It is the norm that everyone tries to adhere to since it satisfies other important organizational goals (customer service, financial gain) without obviously compromising safety.

Outside audits are one way to help an organization break out of the perception that its safety is uncompromised. In other words, neutral observers may better be able to spot the "bad news" among what are normal, everyday decisions and actions to people on the inside.

Finally, when it comes to safety, every organization has room to improve. What separates a strong safety culture from a weak one is not how large this room is. Rather, what matters is whether the organization is willing to explore this space, to find leverage points to learn and improve.

Notes

1. Vaughan, D. (1996). *The Challenger launch decision*. Chicago, IL: University of Chicago Press.
2. The quote on Murphy's law comes in part from Langewiesche, W. (1998). *Inside the sky*. New York: Pantheon.
3. *Airliner World*, 2001, p. 79.
4. *International Herald Tribune*, June 13, 2000.
5. See: Woods, D.D., Johanssen, L.J., Cook, R.I., and Sarter, N. B. (1994). *Behind human error: Cognitive systems, computers and hindsight*. Dayton, OH: CSERIAC, p. 63.
6. Vaughan, D. (1996). *The Challenger launch decision*. Chicago, IL: University of Chicago Press.
7. Wiegmann, D.A., Zhang, H., vonThaden, T., Sharma, G., and Mitchell, A. (2002). *A synthesis of safety culture and safety climate research* (Tech report ARL-02-3/FAA-02-2). University of Illinois at Urbana-Champaign: Aviation Research Lab.

16 Making Recommendations

Coming up with useful recommendations can be difficult. Sometimes only the shallowest of remedies seem to lie within reach. Tell people to watch out a little more carefully. Write another procedure to regiment their behavior. Or just get rid of the particular miscreants altogether. The limitations of such countermeasures are severe and deep, and well-documented:

- People will only watch out more carefully for so long, as the novelty and warning of the mishap wears off;
- A new procedure will at some point clash with operational demands or simply disappear in masses of other regulatory paperwork;
- Getting rid of the miscreants doesn't get rid of the problem they got themselves into. Others always seem to be waiting to follow in their footsteps.

Efforts to understand human error should ultimately point to changes that will truly remove the error potential from a system—something that places a high premium on meaningful recommendations.

Recommendations as Predictions

Coming up with meaningful recommendations may be easier if you think of them as predictions, or as a sort of experiment. Human error is systematically connected to features of the tasks and tools that people work with, and to features of the environment in which they carry out their work. Recommendations basically propose to change some of these features. Whether you want new procedures, new technologies, new training, new safety interlocks, new regulations, more managerial commitment—your recommendations essentially propose to re-tool or re-shape parts of the operational or organizational environment in the hope of altering the behavior that goes on within it.

In this sense your recommendations are a prediction, a hypothesis. You propose to modify something, and you implicitly predict it will have a certain effect on human behavior. The strength of your prediction, of course, hinges on the credibility of the connection you have shown earlier: between the observed human errors and critical features of tasks, tools and environment. With this prediction in hand, you challenge those responsible for implementing your recommendations to go along in your experiment—to see if, over time, the proposed changes indeed have the desired effect on human performance.

The "SMART" acronym

If you want your recommendations to get organizational traction, you may want to consider running them by the following criteria (which constitute the SMART acronym):

- **Specific.** You want your recommendation to be clearly specified. Which parts of what organization do you want to do what and when?
- **Measurable.** You want your recommendation to somehow contain "measurable" criteria for success. This will help you and the organization see if it is indeed implemented, and monitor the effects once it is.
- **Agreed** (or at least agreeable). You want your recommendation to take into account legitimate organizational concerns about production or efficiency, otherwise you will not get any traction with those responsible for implementation and achievement.
- **Realistic.** While your recommendations may be difficult, and challenging to the typical worldviews of those who need to get to work with them, you need to keep them realistic—that is, doable for the person responsible for implementing them.
- **Time-bound.** You may want the implementation of your recommendation to have some kind of suggested expiration date, so decision makers can agree on deadlines for achievement.

High-end or Low-end Recommendations

What kinds of changes can you propose that might have some effect on human performance? A basic choice open to you is how far up a supposed causal chain you want your recommended changes to have an impact.

Typical of reactions to failure is that people start very low or downstream. Recommendations focus on those who committed the error, or on other operators like them. Recommendations low in a causal chain aim for example at retraining individuals who proved to be deficient, or at demoting them or getting rid of them in some other way. Other low-end recommendations may suggest to tighten procedures, presumably regimenting or boxing in the behavior of erratic and unreliable human beings.

Alternatively, recommendations can aim high, at structural decisions regarding resources, technologies and pressures that people in the workplace deal with. High-end recommendations could for example suggest to re-allocate resources to particular departments or operational activities. This choice—upstream or downstream—is more or less yours. And this choice directly influences:

- the ease with which your recommendation can be implemented;
- the effectiveness of your recommended change.

The ease of implementation and the effectiveness of an implemented recommendation generally work in opposite directions. In other words: the easier the recommendation can be sold and implemented, the less effective it will be (see Figure 16.1).

Generally, recommendations for changes low on a causal chain are not very sweeping. They concentrate on a few individuals or a small subsection of an organization. These recommendations are satisfying for people who seek retribution for a mishap, or people who want to "make an example" of those who committed the errors.

But after implementation, the potential for the same trouble is left in place. The error is almost guaranteed to repeat itself in some shape or form, through someone else who finds him- or herself in a similar situation. Low-end recommendations really deal with symptoms, not with causes. After their implementation, the system as a whole has not become much wiser or better.

Remember from Chapter 10: one reason for the illusion that low-end or other narrow recommendations will prevent recurrence is the false idea that failure sequences take a linear path. Take any step along the way out of the sequence, and the failure will no longer occur. In complex, dynamic systems, however, this is hardly ever the case. The pathway towards failure is seldom linear or narrow or simple. Mishaps have dense patterns of causes, with contributions from all corners and parts of the system, and typically depend on many subtle concurrences.

Putting one countermeasure in place somewhere along (what you thought was like) a linear pathway to failure may not be enough. In devising countermeasures it is crucial to understand the vulnerabilities through which

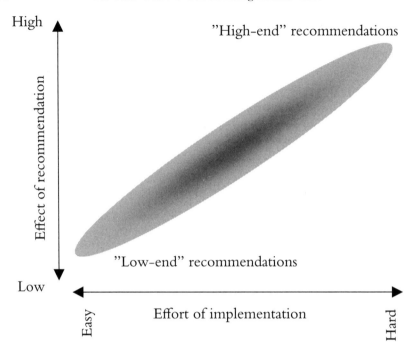

Figure 16.1 The trade-off between recommendations that will be easier to implement and recommendations that will actually have some lasting effect.

entire parts of a system (the tools, tasks, operational and organizational features) can contribute to system failure under different guises or conditions. Recommendations should aim to convert a lesson about a past failure into broader system resilience in the future.

The International Civil Aviation Organization, a United Nations body, identifies three levels of action that decision makers can choose in pursuing safety recommendations.

The first level of action is to eliminate the hazard, thereby putatively preventing a future accident. In the case of a runway collision accident, for example, a decision could be made that in airports having parallel runways, one runway should be used for take-offs and the other for landings. After an accident with in-flight icing of the airframe, it could be decided to absolutely forbid operations when conditions are conducive to airframe icing. These are the safest decisions but they may not be the most efficient.

The second level of action is to accept the hazard identified and adjust the system to tolerate errors and to reduce the possibility of an occurrence. In this context, the decisions following a runway collision might include eliminating intersection take offs or clearances involving taxiing into position on an active runway and holding for take off clearance. In the icing example, the decision might be to eliminate operations into airports without proper de-icing facilities, or when aircraft equipment related to anti-icing protection is unserviceable, in environmental conditions conducive to icing. Although not as safe as first level actions, these options are more realistic and efficient and they could well work.

The third level of action involves both accepting that the hazard can be neither eliminated (level one) nor controlled (level two) and teaching operational personnel to live with it. Typical actions include changes in personnel selection, training, supervision, staffing and evaluation, increasing or adding warnings, and any other modifications which could prevent operational personnel from making a similar mistake. Third level actions should not be taken in preference to first or second level actions, since it is impossible to anticipate all future kinds of problems.

Difficulties with high-end recommendations

The higher you aim in a causal chain, the more difficult it becomes to find acceptance for your recommendation. The proposed change will likely be substantial, structural or wholesale. It will almost certainly be more expensive. And it may concern those who are so far removed from any operational particulars that they can easily claim to bear no responsibility in causing this event or in helping to prevent the next one. Short of saying that it would be too expensive, organizations are good at finding reasons why structural recommendations do not need to be implemented, for example:

- "We already pay attention to that";
- "That's in the manual";
- "This is not our role";
- "We've got a procedure to cover that";
- "This recommendation has no relevance to the mishap";
- "People are selected and trained to deal with that";
- "This is not our problem".

It is easy to be put off before you even begin writing any recommendations. In fact, many recommendations that aim very high in the causal chain do not come out of first investigations, but out of re-opened inquiries, or ones re-submitted to higher authorities after compelling expressions of discontent with earlier conclusions.

One such case was the crash of a DC-10 airliner into Mount Erebus on Antarctica. The probable cause in the Aircraft Accident Report was the decision of the captain to continue the flight at low level toward an area of poor surface and horizon definition when the crew was not certain of their position. The kinds of recommendations that follow from such a probable cause statement are not difficult to imagine. Tighten procedures; exhort captains to be more careful next time around.

A subsequent Commission of Inquiry determined that the dominant cause was the mistake by airline officials who programmed the aircraft computers—a mistake directly attributable not so much to the persons who made it, but to the administrative airline procedures which made the mistake possible. The kinds of recommendations that follow from this conclusion would be different and aim more at the high end. Review the entire operation to Antarctica and the way in which it is prepared and managed. And institute double-checking of computer programming.[1]

The case for including or emphasizing high-end recommendations in a first investigation is strong. If anything, it is discouraging to have to investigate the same basic incident or accident twice. Structural changes are more likely to have an effect on the operation as a whole, by removing or foreclosing error traps that would otherwise remain present in the system.

Judge Moshansky's investigation of the Air Ontario crash generated 191 recommendations. Most of these were high-end. They concerned for example:[2]

- *Allocation of resources to safety versus production activities;*
- *Inadequate safety management by airline and authority alike;*
- *Management of organizational change;*
- *Deficiencies in operations and maintenance;*
- *Deficient management and introduction of new aircraft;*
- *Deficient lines of communication between management and personnel;*
- *Deficient scheduling (overcommitting this particular aircraft);*
- *Deficient monitoring and auditing;*
- *Deficient inspection and control and handling of information;*
- *Inadequate purchasing of spares;*
- *Low motivation and job instability following airline merger;*
- *Different corporate cultures;*
- *High employee turnover;*
- *Poor support to operational personnel;*
- *Inadequate policy making by airline and authority.*

These are just some of the areas where recommendations were made. With serious efforts to understand human error, many of these kinds of conditions can probably be uncovered in any complex system. The ability to generate structural recommendations that aim high up in a causal chain is a reflection of the quality and depth of your understanding of human error.

Searching the Evidence for Countermeasures

The kind and content of your recommendations depends, of course, on the kind and content of the events you are trying to understand. But inspiration for high-end recommendations may come from some of the organizational contributions from previous chapters:

- The re-allocation of resources that flow from the blunt end, and the alleviation of constraints that are imposed on operators' local decisions and trade-offs;
- Making goal conflicts explicit and turning them into topics for discussion among those involved;
- Make regulatory access more meaningful through a re-examination of the nature and depth of the relationship between regulator and operator.

Get help from the participants

If possible, it can be fruitful to build on the list above by talking to the participants themselves. These are some of the questions that Gary Klein (see Chapter 11) and his researchers ask participants when looking for countermeasures against recurrence of the mishap:

- What would have helped you to get the right picture of the situation?
- Would any specific training, experience, knowledge, procedures or cooperation with others have helped?
- If a key feature of the situation would have been different, what would you have done differently?
- Could clearer guidance from your company have helped you make a better trade-off between conflicting goals?

Not only can answers to these questions identify countermeasures. They can also serve as a reality check. Would the countermeasures you think about proposing have any effect on the kind of situation you are trying to avoid? Asking the participants themselves, who may have intimate knowledge of the situation, may be a good idea.

Persistent Problems with Recommendations

Investigations are typically expensive. Organizations can end up allocating significant resources to probing an incident, regardless of whether they want to or have to because of regulations. The money spent on an investigation ups the stakes of the recommendations that come out of it. Apart from producing and circulating incident stories that may have a learning benefit (see Chapter 15), recommendations are the only product to have some lasting effect in how an organization and its activities are designed or run. If all recommendations are rejected at the end, then the investigation has missed its goal.

One investigator described how the writing and inclusion of recommendations is heavily determined by who is going to be on the committee assessing the recommendations for implementation. Language may be adjusted or changed, some recommendations may be left out in order to increase the chances for others. This illustrates that the road from investigation to implementation of countermeasures is largely a political one.

The example above also reminds us of the disconnect between the quality of an investigation and its eventual impact on organizational practice. A really good investigation does not necessarily lead to the implementation of really good countermeasures. In fact, the opposite may be true if you look at Figure 16.1. Really good investigations may reveal systemic shortcomings that necessitate fundamental interventions which are too expensive or sensitive to be accepted.

The focus on diagnosis, not change

Recommendations represent the less sexy, more arduous back-end of an investigation. One reason why they can fall by the wayside and not get implemented with any convincing effect is that they are fired off into an organization as limited, one-shot fixes. Many organizations—even those with mature safety departments and high investments in investigations—lack a coherent strategy on continuous improvement. Resources for quality control and operational safety are directed primarily at finding out what went wrong in the past, rather than assuring that it will go right in the future. The focus is on diagnosis, not change.

The emphasis on diagnosis can hamper progress on safety. Recommendations that are the result of careful diagnosis have not much hope of succeeding if nobody actively manages a dedicated organizational improvement process. Similarly, feedback about the success of implemented recommendations will not generate itself. It needs to be actively sought out, looked for, compiled and sent back and assessed. Organizations seldom have mechanisms in place for generating feedback about implemented recommendations, and enjoy little support or understanding from senior management for the need for any such mechanisms. The issue, in the end, is one of sponsoring and maximizing organizational learning; or about making necessary organizational changes. This is what the next chapters are all about.

Notes

1   See: Vette, G. (1983). *Impact Erebus*. Auckland, NZ: Hodder and Stoughton.
2   Moshansky, V.P. (1992). *Commission of inquiry into the Air Ontario accident at Dryden, Ontario* (Final report, vol. 1–4). Ottawa, ON: Minister of Supply and Services, Canada.

17 Abandon the Fallacy of a Quick Fix

So how do you make a human error problem go away? The answer isn't as simple as the question.

Your human error problem is an organizational problem. Your human error problem is at least as complex as the organization that helped create it. If you want to do something about your human error problem, you will have to start seeing it as the effect of problems deeper inside your organization. Not as the simple cause of all your trouble. Once organizations really start fretting over the question of how to make a human error problem go away, there could be a window in which there is more openness to learning:

- Parts of an organization may welcome self-examination more than before;
- Traditional lines between management and operators, between regulators and operators, may be temporarily blurred in joint efforts to find out what went wrong and why;
- People and the systems they work in may be open to change—even if only for a short while;
- Resources may be available that are otherwise dedicated to production only, something that could make even the more difficult recommendations for change realistic.

Of course, this atmosphere of openness, of willingness and commitment to learn and improve, can quickly become compromised by calls for accountability, by primitive knee-jerk reactions toward putative miscreants. Even the formal process of investigating mishaps and coming up with recommendations for change may itself stand in the way of learning from failure. In the aftermath of failure, the pressure to come up with findings and recommendations quickly can be enormous—depending on the visibility of the industry or the accident. An intense concern for safety (or showing such concern) can translate into pressure to reach closure quickly, something that can lead to a superficial study of the mishap and Old View countermeasures.

The Political Landscape of Learning

Learning is about:

- Identifying and acknowledging the real sources of operational vulnerability;
- Modifying an organization's basic assumptions and beliefs about what makes their operations safe or risky;
- Creating leverage over organizational decisions that affect the real sources of operational resilience and vulnerability.

These three are not always (in some industries not often) welcome activities. Some industries actively discourage them altogether. Mishaps, and efforts to turn an organization around to the New View, always expose fault lines and can inflame existing rifts. They challenge commonly held beliefs about what makes the organization safe or risky. And you may be asking people to stop mobilizing their favorite, simplistic explanatory mechanisms ("human error!") and associated countermeasures ("punish the culprit!"). Indeed, not learning is about:

- Keeping the real sources of operational vulnerability out of sight, so that the organization does not have to deal with the expensive, fundamental implications of failure;
- Not modifying an organization's basic assumptions and beliefs about what makes their operations safe or risky. This allows Bad Apple explanations to persist;
- Refusing leverage over organizational decisions that affect the real sources of operational resilience and vulnerability, so that goals other than safety can keep on taking precedence.

Efforts to uncover deeper sources of operational vulnerability could make some people uncomfortable, depending on their histories and organizational responsibilities. Those sources of operational vulnerability may have a lot to do with their work, and less with the operators whose errors finally give expression to the problems created. For example:

- Resources for training may have been trimmed. People may have been rushed through training programs or promotions to fill a production quota;
- An obvious managerial commitment to safety may be lacking. The resources made available for safety efforts could be unknown. A safety policy may not exist;

- Personnel may have been cut, requiring fewer people to do the same job;
- Production pressure may have increased;
- People may have been given additional roles in addition to their core jobs. This not only generates more production pressure but also adds a sense of role ambiguity.

Such measures often reflect legitimate organizational trade-offs that take economic concerns seriously (and may help the organization survive in the first place). But you must not misinterpret the inevitable downstream operational problems that result from an organization trying to do more with less. These signs can, and sometimes will, be interpreted as local "human errors". But errors are consequences: the leakage that occurs around the edges when you put pressure on a system without taking other factors into account.

Of course, the local rationality principle applies to managers as much as it does to other people. What managers were doing at the time made sense given their pressures, understanding of the situation, goals and knowledge. Managers' ability to embrace New View countermeasures then, hinges on these "making sense" to them—given their goals, knowledge and focus of attention.

Supposed Quick Fixes

Here are some really obvious, but false, quick fixes:

- Retraining the people involved in a mishap. Who says it's just them? Their performance could be symptomatic of something deeper;
- Reprimanding the people involved. Your organization cannot learn and punish at the same time;
- Writing a procedure. This often only deals with the latest hole uncovered by the mishap, and simplistically assumes linear accident trajectories. More procedures also create more intransparency and non-compliance;
- Adding just a little bit more technology. More technology will create new work for people, new error opportunities and pathways to breakdown.

So much for these, then. There are other, more subtle fixes that actually fix very little and obstruct a lot. These efforts can look like learning and may keep political constituencies, bureaucratic hierarchies or media representatives happy. But if you really want to understand human error, this is not the way to go:

"To err is human"

Although it is a seemingly forgiving stance to take, organizations that suggest that "to err is simply human" may normalize error to the point where it is no longer interpreted as a sign of deeper trouble.

"There is one place where doctors can talk candidly about their mistakes. It is called the Morbidity and Mortality Conference, or more simply, M. & M. Surgeons, in particular, take M. & M. seriously. Here they can gather behind closed doors to review the mistakes, complications and deaths that occurred on their watch, determine responsibility, and figure out what to do differently next time."

A sophisticated instrument for trying to learn from failure, M. & M.s assume that every doctor can make errors, yet that no doctor should—avoiding errors is largely a matter of will. This can truncate the search for deeper, error-producing conditions. In fact, "the M & M takes none of this into account. For that reason, many experts see it as a rather shabby approach to analyzing error and improving performance in medicine. It is isn't enough to ask what a clinician could or should have done differently so that he and others may learn for next time. The doctor is often only the final actor in a chain of events that set him or her up to fail. Error experts, therefore, believe that it's the process, not the individuals in it, which requires closer examination and correction."[1]

"Setting examples"

Organizations that believe they have to "set an example" by punishing or reprimanding individual operators are not learning from failure. The illusion is there, of course: if error carries repercussions for individuals, then others will learn to be more careful too.

The problem is that instead of making people avoid errors, an organization will make people avoid the reporting of errors, or the reporting of conditions that may produce such errors.

In one organization it is not unusual for new operators to violate operating procedures as a sort of "initiation rite" when they get qualified for work on a new machine. By this they show veteran operators that they can handle the new machine just as well. To be sure, not all new operators take part, but many do. In fact, it is difficult to be sure how many take part. Occasionally, news of the violations reaches management, however. They respond by punishing the individual violators (typically demoting them), thus "setting examples".

The problem is that instead of mitigating the risky initiation practice, these organizational responses entrench it. The pressure on new operators is now not only to violate rules, but to make sure that they aren't caught doing it—making the initiation rite even more of a thrill for everyone. The message to operators is: don't get caught violating the rules. And if you do get caught, you deserve to be punished—not because you violate the rules, but because you were dumb enough to get caught.

A proposal was launched to make a few operators—who got caught violating rules even more than usual—into teachers for new operators. These teachers would be able to tell from their own experience about the pressures and risks of the practice and getting qualified. Management, however, voted down the proposal because all operators expected punishment of the perpetrators. "Promoting" them to teachers was thought to send entirely the wrong message: it would show that management condoned the practice.

Compartmentalization

One way to deal with information that threatens basic beliefs and assumptions about the safety of the system is to compartmentalize it; to contain it.

In the organization described above, the "initiation rite" takes place when new operators are qualifying for working on a new machine. So, nominally, it happens under the auspices of the training department. When other departments hear about the practice, all they do is turn their heads and declare that it is a "training problem". A problem, in other words, of which they have no part and from which they have nothing to learn.

While this may look like a palatable solution (give the problem to the experts and let them deal with it!), compartmentalization limits the reach of safety information. The assumption beneath compartmentalization is that the need to change—if there is a need at all—is an isolated one: it is someone else's problem. There is no larger lesson to be learned (about culture, for example) through which the entire organization may see the need for change. In the example above, were not all operators—also all operators outside the training department—once new operators, and thus maybe exposed to or affected by the pressures that the initiation rite represents?

What seems to characterize high reliability organizations (ones that invest heavily in learning from failure) more than anything is the ability to identify commonalities across incidents. Instead of departments distancing themselves from problems that occur at other times or places and focusing on the differences and unique features (real or imagined), they seek similarities that contain lessons for all to learn.

Creating safety departments or safety professionals can have another side effect too. Some safety professionals have become so divorced from daily operations that they only have a highly idealized view of the actual work processes. They are no longer able to identify with the point of view of people who actually do the safety critical work every day.

Management says "we did it too"

What characterizes many safety-critical organizations is that senior managers were often operators themselves—or still are (part-time). For example, in hospitals, physicians run departments, in airlines pilots do. On the one hand this provides an opportunity. Managers can identify with operators in terms of the pressures and dilemmas that exist in their jobs, thus making it easier for them to get access to the underlying sources of error.

But it can backfire too. The fact that managers were once operators themselves may rob them of credibility when it comes to proposing fundamental changes that affect everyone.

The organization in the examples above is one where senior management is made up of operators or ex-operators. What if management would want to reduce the risks associated with the initiation practice, or eliminate it altogether? They were once new operators themselves and very likely did the same thing when getting qualified. It is difficult for them to attain credibility in any proposal to curb the practice.

Blaming someone else: the regulator for example

Most safety-critical industries are regulated in some way. With the specific data of a mishap in hand, it is always easy to find gaps where the regulator "failed" in its monitoring role. This is not a very meaningful finding, however. Identifying regulatory oversights in hindsight does not explain the reasons for those—what now look like—obvious omissions. Local workload, the need to keep up with ever-changing technologies and working practices and the fact that the narrow technical expertise of many inspectors can hardly foresee the kinds of complex, interactive sequences that produce real accidents, all conspire against a regulator's ability to exercise its role.

If you feel you have to address the regulator in your investigation, do not look for where they went wrong. As with investigating the assessments and actions of operators, find out how the regulator's trade-offs, perceptions and judgments made local sense at the time; why what they were doing or looking at was the right thing given their goals, resources, and understanding of the situation.

Another complaint often leveled against regulators is that they collude with those they are supposed to regulate, but this is largely a red herring (and, interestingly, almost universally disagreed with by those who are regulated.

Independent of claims to collusion, they often see regulators as behind the times, intrusive and threatening). To get the information they need, regulators are to a large extent dependent on the organizations they regulate, and likely even on personal relationships with people in those organizations. The choice, really, is between creating an adversarial atmosphere in which it will be difficult to get access to safety-related information, or one in which a joint investment in safety is seen as in everybody's best interest.

As soon as you "pass the buck" of having to learn from a mistake to someone else—another person, another department, another company (e.g. suppliers), another kind of organization (e.g. regulator versus operator), you are probably shortchanging yourself with respect to the lessons that are to be learned. Failure is not someone else's problem.

The Hard Fixes

So what should you expect or ask your organization to do instead? The quick fixes may be no good, and may in fact be a sign of your organization trying not to learn. But what, then, are the hard fixes? Hard fixes are evidence of people inside the organization taking the surprise of a failure seriously. Rather than trying to reduce that surprise by pinning the responsibility for failure on a few local miscreants, these people will understand that failure tries to tell them something real about their organization and how it has managed risk and produced safety so far. The failure may show, for example:

- How otherwise legitimate trade-offs between safety and goals such as production or efficiency may have made it increasingly difficult for people to create safety through practice;
- How the entire system may have drifted towards the boundaries of safe performance, taking the definition of "acceptable risk" or acceptable practice along with it;
- How incentive structures governing people's performance may actually have encouraged a focus on economic or efficiency goals, with safety being taken for granted;
- How the priorities and preferences that employees express through their own practice may be a logical reproduction of that which the entire organization finds important;
- How managerial attention to safety has eroded, or never really developed, leaving the organization without explicit safety policy and no clarity about the resources committed to safety;

- How organizational models about the sources of risk, and organizational plans for how to deal with risk, may have been wrong or incomplete.

Hard fixes change something fundamental about the organization. This is what makes them hard. But this is also what makes them real fixes.

What you should do depends on how safe you already are

Which hard fixes you can successfully deploy depends on how safe your system already is. It makes an enormous difference if you want to make mountain climbing safer or improve the safety of scheduled airline operations. The difference lies partly in how safe these activities already are, and what they already have in place to help make their practices safe. René Amalberti has identified four different types of systems, each of increasing safety levels:[2]

- **Unsafe systems**, such as certain types of mountain climbing or surgery (e.g. transplant). The risk of failure (including fatalities) is inherent in the activity and is accepted, as it is the other side of trying to extract maximum performance, or simply worth the bet because the alternative is even worse. Outstanding performance and a constant search for maximum performance make unsafe systems work. Greater safety can sometimes be expected through increasing practitioner competence. That often takes care of itself, though, as participants in such activities tend to be competitive and constantly out to improve their own performance.
- **Safer systems**, such as road traffic or certain types of healthcare. Safety improvements here can come from standardization—of participants (through training), of the work (through rules and procedures) and of the technology used (through ergonomics). People in safer systems make their own choices, but such choices can be led in better ways through standardization. Quality control can help monitor whether components and processes meet prespecified standard criteria. (These hard fixes may be tried in unsafe systems as well, but they could well be resisted as so much comes down to an expression of individual competence there.)
- **Safe systems**, such as our food supply or charter airline flying. Safety in these systems comes in part from top management commitment to safety. This commitment is made obvious by a written safety policy and an explicit advertisement of how much resources management spends on safety. The organization does safety monitoring beyond quality control (e.g. through event reporting and deeper analysis), safety management (e.g. by safety

departments), and emphasizes skills that go beyond competent individual practice (e.g. teamwork or other "soft" skills). In these safe systems, incident reporting is worthwhile, as the ingredients of incidents (dress rehearsals) tend to show up in accidents.

A written safety policy can be helpful in documenting and showing managerial commitment to safety, and help standardize practices and safety initiatives. Safety policies generally begin with a statement of intent that outlines the organization's overall philosophy in relation to the management of safety, including reference to the broad responsibilities of both management and workforce.

Safety policies often also detail people and their duties in relation to safety. It outlines the chain of command in terms of safety management, saying who is responsible for what and how implementation of the policy is managed. It should also specify that safety management responsibilities rank equal in position with responsibilities for cost and production (few organizations actually do this).

Finally, safety policies detail the practical arrangements through which the policy actually gets implemented (how management should take action on safety problems, communicate about safety, deal with incidents, safety training, and so forth).

- **Ultra-safe systems**, such as scheduled international airline flying or nuclear power generation. In addition to the safety measures taken by the lower levels, a key feature here is safety supervision. Automatic monitoring of process parameters and automatic (computerized) execution of many routine tasks helps compliance and predictability. The route to risk in ultra-safe systems is one where the entire system gradually changes its definition of what is safe or acceptable practice. The slow drift into failure is the biggest source of residual risk here. This means that incident reporting (still effective at the previous level) may no longer help predict or prevent accidents, as the definition of an "incident" has shifted for everybody, including supervisory bodies (e.g. regulators). Accident ingredients will not have been reported as incidents, for they weren't seen as such at the time.

It would seem that improving the safety of your system involves pegging that system somewhere along Amalberti's four levels. And then getting inspiration from the next level to find out what to do and where to go. So if parts of your practices are unsafe, then more standardization of equipment and work practices could help. And if you have a safe system and seem to have extracted the maximum return from event reporting, you may need to try t0 fix your system on higher-order variables in addition to that (see the next section).

In any case, you should probably encourage your system to keep on doing what it has been doing (e.g. emphasize individual competence, or do event reporting), but turn to measures from the next level if you want to make your system even safer.

Yet it may not be simple to get traction with such proposals. While the various forms of safety management are, in principle, available to everybody, industries apply them quite variously. This is interesting. It suggests that systems do not spontaneously try to move up from one level of safety to the next. There is not a constant pressure for an upward flow. Indeed, a safety level may be a societally acceptable balance between having a safe system, and having that system at all (as it may not otherwise be economically viable). This means that systems may not automatically be open for inspiration from the next safety level up. They may be content with their safety level, and with the safety measures already taken at that level.

How to be Confident that your System can get Safer

So how do you know whether your efforts will have any effect? Here are some key questions you should ask yourself and your organization:

- **Monitoring of safety monitoring** (or meta-monitoring). Does your organization invest in an awareness of the models of risk it embodies in its safety strategies and risk countermeasures? Is it interested to find out how it may have been ill-calibrated all along, and does it acknowledge that it needs to monitor how it actually monitors safety? This is important if your organization wants to avoid stale coping mechanisms, misplaced confidence in how it regulates or checks safety, and does not want to miss new possible pathways to failure.
- **Past success as guarantee of future safety**. Does your organization see continued operational success as a guarantee of future safety, as an indication that hazards are not present or that countermeasures in place suffice? In this case, its ability to deal with unexpected events may be hampered. In complex, dynamic systems, past success is no guarantee of continued safety.
- **Distancing through differencing.** In this process, organizational members look at other failures and other organizations as not relevant to them and their situation. They discard other events because they appear to be dissimilar or distant. But just because the organization or section has different technical problems, different managers, different histories, or can claim to already have

addressed a particular safety concern revealed by the event, this does not mean that they are immune to the problem. Seemingly divergent events can represent similar underlying patterns in the drift towards hazard.
- **Fragmented problem solving**. It could be interesting to probe to what extent problem-solving activities are disjointed across organizational departments, sections or subcontractors, as discontinuities and internal handovers of tasks increase risk. With information incomplete, disjointed and patchy, nobody has the big picture, and nobody may be able to recognize the gradual erosion of safety constraints on the design and operation of the original system that move an organization closer to the edge of failure.
- **Knowing the gap between work-as-imagined and work-as-done.** One marker of resilience is the distance between operations as management imagines they go on and how they actually go on. A large distance indicates that your organizational leadership may be ill-calibrated to the challenges and risks encountered in real operations. Also, they may also miss how safety is actually created as people conduct work, construct discourse and rationality around it, and gather experiences from it.
- **Keeping the discussion about risk alive** even (or especially) when everything looks safe. One way is to see whether activities associated with recalibrating models of safety and risk are going on at all. This typically involves stakeholders discussing risk even when everything looks safe. Indeed, if discussions about risk are going on even in the absence of obvious threats to safety, you could get some confidence that your organization is investing in an analysis, and possibly in a critique and subsequent update, of its models of risk.
- **Having a person or function within the system with the authority, credibility and resources** to go against common interpretations and decisions about safety and risk. Historically, "whistleblowers" may hail from lower ranks where the amount of knowledge about the extent of the problem is not matched by the authority or resources to do something about it or have the system change course. Your organization shows a level of maturity if it succeeds in building in this function at meaningful organizational levels. This also relates to the next point.
- **The ability and extent of bringing in fresh perspectives**. Organizations that apply fresh perspectives (e.g. people from other backgrounds, diverse viewpoints) on problem-solving activities seem to be more effective in managing risk: they generate more hypotheses, cover more contingencies, openly debate rationales for decision making, reveal hidden assumptions. With a neutral observer or commentator thus "institutionalized", you could

be slightly more confident that your organization can help "self-regulate" its own safety.

These questions can be helpful in organizations at any level of safety. But they have particular value for ultra-safe systems where all the low-hanging fruit has already been picked. Where the defining risk is a drift into failure, an ability to question and update models of risk (and what is acceptable practice) is crucial.

Notes

1   Gawande, A. (1999). When doctors make mistakes. *The New Yorker*, February 1, pp. 40–55.
2   Amalberti, R. (2006). Optimum system safety and optimum system resilience: Agonistic or antagonistic concepts? In E. Hollnagel, D.D. Woods, N.G. Leveson (eds). *Resilience engineering: Concepts and precepts*. Aldershot, UK: Ashgate Publishing Co., pp. 238–256.

18 What about People's Own Responsibility?

Remember the tension from Chapter 1? There are always two ways to construct a story about human error:

- You can build a story about individuals, about sharp end operators. These people made errors. They did not follow the rules; they did not try hard enough. They should have done something else. But this first, obvious story, is always incomplete and probably wrong in many ways. It contains counterfactuals, and typically offers judgments instead of explanations.
- The second, deeper story, in contrast, is a story about the organization surrounding people at the time; about the system that accompanied and helped produce the behavior at the sharp end. When you relocate people's behavior in this setting, when you put it back into context, it is likely not at all so surprising, so unexpected, so deviant. What people did made sense to them at the time, given features of their work, their tools and tasks and surrounding organization. It would probably have made sense to others too.

Indeed, as soon as you have reason to believe that any other practitioner would have done the same thing as the one whose assessments and actions are now controversial, you should start looking at the system. For it is the system surrounding people, then, that may play a large role in producing the observed behavior. We call this the "substitution test". Substitute one practitioner for another (even as a thought experiment) and imagine if the same thing would have happened. In that case, start probing deeper.

But somewhere in going from the first to the second story, in migrating from the Bad Apple Theory to the New View, don't you lose sight of the individual and his or her responsibility? Should you not hold people accountable, in some way, for the mistakes that they are involved in? That is what this chapter is about.

Getting people off the hook?

The Field Guide may be, to some, a conspiracy to get culprits off the hook. It may amount, to some, to one big exculpatory exercise—explaining away error; making it less bad; less egregious; less sinful; normalizing it.

The reaction is understandable. Some people always need to bear the brunt of a system's failure. These are often people at the blunt end of organizations. Managers, supervisors, boardmembers: *they* have to explain the failure to customers, patients, passengers, stockowners, lawyers. The Field Guide may look like a ploy to excuse defective operators; to let them plod on with no repercussions, to leave them as uncorrected unreliable elements in an otherwise safe system. But the Field Guide aims to help you explain the riddle of puzzling human performance—not explain it away. There is a difference between explaining and excusing human performance. The former is what the Field Guide helps you do.

No Responsibility without Proof of Authority

As soon as you start probing the surrounding setting or organization for explanations of failure, you inevitably raise questions about where the responsibility of people ends and that of context begins. Of course you need to consider the accountability of practitioners, especially practitioners who have jobs where they are reponsible for creating safety for other people. Actually, most practitioners even want, and expect, accountability. It gives their jobs meaning. The possibility of culpability is the other side of the feeling of control that their work otherwise gives to them.

Physician Atul Gawande comments on a recent surgical incident and observes that terms such as "systems problems" are part of a "dry language of structures, not people". He then goes on to say how "something in me, too, demands an acknowledgement of my autonomy, which is also to say my ultimate culpability ... although the odds were against me, it wasn't as if I had no chance of succeeding. Good doctoring is all about making the most of the hand you're dealt, and I failed to do so."[1]

But you cannot ask somebody to be responsible for something he or she had no control over. It is impossible to hold somebody accountable for something over which that person had no authority. And that is exactly where the problem lies. This is at the heart of the conflict between the Old View and the New

View. You can define accountability as responsibility for which the person in question had requisite authority:

$$\text{Accountability} = \text{Responsibility} + \text{Requisite Authority}$$

Responsibility-authority mismatches

Real work is full of reponsibility-authority mismatches. Where people have formal responsibility for the outcome of their work, but do not have full authority over the actions and decisions that take them to that outcome. The question is not whether the work of the people whose errors you want to understand contained these mismatches (because it most likely did). The question is how *you* deal with them:

- The Old View does not take responsibility-authority mismatches seriously. It glosses over them. It implies that people should take responsibility for the outcomes of their actions simply because their formal status demands it (Four stripes, a Medical Degree) and their paycheck and societal standing compensates for it. If they don't like that, they should not have been in that job.
- The New View does take responsibility-authority mismatches seriously. Not just for considering how "accountable" you can actually claim anybody was, but also because the existence and extent of responsibility-authority mismatches tells you something essential about the organization. Does management, for example, acknowledge such mismatches on part of its personnel (or itself, for that matter)? Does it try to address them in any meaningful way?

"This is at the heart of the professional pilot's eternal conflict", writes Wilkinson in a comment to the November Oscar case (see Chapter 1). *"Into one ear the airlines lecture, 'Never break regulations. Never take a chance. Never ignore written procedures. Never compromise safety.' Yet in the other they whisper, 'Don't cost us time. Don't waste our money. Get your passengers to their destination—don't find reasons why you can't'."*[2]

The responsibility-authority mismatch brings us back to the basic goal conflicts that drive most safety-critical and time-critical work. Such work consists of holding together a tapestry of multiple competing goals, of reconciling them as best as possible in real-time practice.

As a result, the work involves the sacrificing decisions that the previous chapter talked about: sacrificing safety for efficiency, reliability for cost reduction, diligence for higher production. It involves efficiency-thoroughness trade-offs (ETTOs), as Erik Hollnagel calls them. If an entire system is crying out for operators to be efficient, how can you then turn around after the occasional failure and all of a sudden demand that they should have been thorough all along instead? Of all the unreasonable things that we wreak upon one another in the wake of failure, says Erik Hollnagel, this is among the most unreasonable.[3]

It does lay down the golden rule. Holding people accountable is fine. But you need to be able to show that people had the authority to live up to the reponsibility that you are now asking of them. If you can't do that, your calls for accountability have no merit.

How authority gets constrained

Proving that people had full authority over the situation for which you now want to hold them responsible is very difficult. Full authority is almost always constrained. For example:

- Operational decisions can (and almost always do) have an economic impact which can be hard to oversee for practitioners themselves. As a result, practitioners may need to consult with other experts first (e.g. accountants, insurance people), blurring their *de facto* (if not formal) authority.
- Operational decisions almost always need to be made within limited time and under conditions of uncertainty. Attaining full knowledge of the decision space and its possible consequences is impossible.
- The ability to recruit additional operational expertise from outside the situation may be hampered (they are out there alone, flying over the ocean, hovering above the operating table, sitting at the radar scope). This limits people's ability to form a full or alternative picture of the situation.

These, and other, aspects of safety-critical work limit the authority that operational people actually have over their work and its possible outcomes.

ValuJet flight 592 crashed after take-off from Miami airport because oxygen generators in its cargo hold caught fire. The generators had been loaded onto the airplane by employees of a maintenance contractor, who were subsequently prosecuted. The editor of Aviation Week and Space Technology "strongly believed the failure of SabreTech employees to put caps on oxygen generators constituted willful negligence that led to the killing of 110 passengers and crew. Prosecutors were right to bring charges. There has to be some fear that not doing one's job correctly could lead to prosecution."[4]

But holding individuals accountable by prosecuting them misses the point. It shortcuts the need to learn fundamental lessons, if it acknowledges that fundamental lessons are there to be learned in the first place. In the SabreTech case, maintenance employees inhabited a world of boss-men and sudden firings, and that did not supply safety caps for expired oxygen generators. The airline may have been as inexperienced and under as much financial pressure as people in the maintenance organization supporting it. It was also a world of language difficulties—not only because many were Spanish speakers in an environment of English engineering language:

"Here is what really happened. Nearly 600 people logged work time against the three ValuJet airplanes in SabreTech's Miami hangar; of them 72 workers logged 910 hours across several weeks against the job of replacing the 'expired' oxygen generators—those at the end of their approved lives. According to the supplied ValuJet work card 0069, the second step of the seven-step process was: 'If the generator has not been expended install shipping cap on the firing pin.'

This required a gang of hard-pressed mechanics to draw a distinction between canisters that were 'expired', meaning the ones they were removing, and canisters that were not 'expended', meaning the same ones, loaded and ready to fire, on which they were now expected to put nonexistent caps. Also involved were canisters which were expired and expended, and others which were not expired but were expended. And then, of course, there was the simpler thing—a set of new replacement canisters, which were both unexpended and unexpired."[5]

These were conditions that existed long before the ValuJet accident, and that exist in many places today. Fear of prosecution stifles the flow of information about such conditions. And information is the prime asset that makes a safety culture work. A flow of information earlier could in fact have told the bad news. It could have revealed these features of people's tasks and tools; these long-standing vulnerabilities that form the stuff that accidents are made of. It would have shown how human error is inextricably connected to how the work is done, with what resources, and under what circumstances and pressures.

At first sight, it is so easy to claim that the individuals in question should have tried a little harder, should have looked a little better, should have been more motivated, concentrated. But on closer inspection, you quickly discover a context that conspired, in various obvious and less obvious ways, against people's ability to do a good job. Even if they came to work to do a good job, the definition of a "good" job may have shifted towards production and punctuality, towards customer service and efficiency, towards beating the competition. This happens in the typically incremental, drifting fashion that is hard to notice.

Holding People Accountable without Proof of Authority

Holding people accountable without proving that they were actually in control over the situation happens a lot. The Old View lives by it, actually. As said in Chapter 1, it satisfies the immediate urges that bubble up in the wake of failure. At least you can show that you are doing something about it! At least you are holding somebody accountable! But just holding people accountable has negative consequences—inevitably. It makes it more difficult, not to say impossible, for your organization to learn from failure. Here is why:

- Just holding people accountable is about keeping your beliefs in a basically safe system intact. By hanging out a few individual miscreants, you can sustain your hope that the only threat comes from some unreliable people. In contrast, learning from failure involves changing these beliefs, and changing the system accordingly. Your problem is not unreliable people in an otherwise safe system. Improving safety comes first from abandoning efforts to repress errors.
- Just holding people accountable is about seeing the culprits as unique parts of the failure, as in: it would not have happened if it were not for them. Learning is about seeing the failure as a part of the system, as produced by the system.
- Just holding people accountable is about teaching your people not to get caught the next time. Learning is about countermeasures that remove error-producing conditions so there won't be a next time.
- Just holding people accountable is about stifling the flow of safety-related information (because people do not want to get caught). Learning is about increasing that flow.
- Just holding people accountable is about closure, about moving beyond the terrible event. Learning is about continuity, about the continuous improvement that comes from firmly integrating the terrible event in what the system knows about itself.

Just holding people accountable misses the point entirely. It zooms the story in on one, or a few, individuals who should have done something else. The deeper complexity (expended, unexpended, expired, unexpired, no caps) is missed or discounted.

Hunting down individuals stifles the flow of information about these sorts of conditions that expose systems to risk. And information is the prime asset that makes a safety culture work. A flow of information earlier could often have told the bad news. It could have revealed the features of people's tasks and tools, the long-standing vulnerabilities that form the stuff that accidents are made of. It would have shown how human error is inextricably connected to how the work is done, with what resources, and under what circumstances and pressures.

Nature or Nurture?

In a way, the tension between Old View and New View explanations of failure comes down to an age-old (and unresolved) debate within psychology. Nature or nurture?

- Is human failure part of our nature? Are people born sinners in otherwise safe systems?
- Or is it a result of nurture? Are people saints at the mercy of deficient organizations, not quite holding their last stand against systems full of latent failures?

People are neither just sinners nor simply saints—at best they are both. The debate, and the basic distinction in it, oversimplifies the challenges before you if you really want to understand the complex dynamics that accompany and produce human error. Safety (and failures) are emerging features; they are by-products of entire systems and how these function in a changing, uncertain and resource-constrained world. Neither safety nor failure are the prerogative of select groups or individuals inside of these systems.

New Models of Accountability

Accountability doesn't need to mean holding people responsible in the traditional sense. You can think about ways to "hold people to account" without invoking all kinds of psychological and political defense mechanisms. As soon as you put people on the defensive, just imagine what happens to possibilities for learning from failure. They disappear. People will cover up, not tell you things, change or leave out inconvenient details.

Pushing accountability up the chain of command

Suppose you don't want to hold low-level operators accountable; you don't make the little guy carry the entire explanation of a mishap on his back. Can you instead direct calls for accountability to higher organizational levels? Is there any positive pay-off in getting managers or directors to know they will be held accountable if things go wrong? Will this make them think twice in where they put their priorities? Will this have any rehabilitative or example-setting effect?

Don't get your hopes up. There is actually evidence that people, at all levels in an organization, will invest in defensive countermeasures, when you think you have to "make them pay" for mistake and disaster. They will start ducking the debris, they will think foremost about their own careers, their own lives. The common good of having the entire organization learn from a mishap gets sacrificed for personal protection and immunization.

Let them tell their story

The real key is holding people accountable without invoking defense mechanisms.

I was visiting the chief executive of a large organization to talk about safety when news came in about an incident that had just happened in their operation. A piece of heavy equipment, that should have been fastened, came loose and caused quite a bit of damage. As I was sitting there, the first reaction around the board table was 'Who did this?! We must get our hands on this person and teach him a real lesson! We should turn him into an example for others! This is unacceptable!'

After the situation had calmed a bit, I suggested that if they really wanted other people to learn from this event, then it could be more profitable to talk to the person in question and ask him to write up his account of what happened and why. And then to publish this account widely throughout the company. If the person would go along with this, then management should drop all further calls for accountability or retribution. It took some effort, but eventually they seemed to agree that this could be a more meaningful way forward.

What is the moral of this encounter?

- If you hold somebody accountable, that does not have to mean exposing that person to liability or punishment;

- You can hold people accountable by letting them tell their story, literally "giving their account";
- Storytelling is a powerful mechanism for others to learn vicariously from trouble.

Many sources point to the value of storytelling in preparing operators for complex, dynamic situations in which not everything can be anticipated. Stories contain valuable lessons about the kinds of trade-offs and sacrificing decisions that, after-the-fact, can be construed as controversial (people were efficient rather than thorough). Stories are easily remembered, scenario-based plots with actors, intentions, a narrative arc, and outcomes that in one way or another can be mapped onto current difficult situations and matched for possible ways out. Incident-reporting systems can capitalize on this possibility. In contrast, more incriminating forms of accountability actually retard this very quality. It robs from people the incentive to tell stories in the first place.

Notes

1   Gawande, A. (2002). *Complications: A surgeon's notes on an imperfect science.* New York: Picado, p. 73.
2   Wilkinson, S. (1996). The November Oscar incident. *Smithsonian Air and Space*, February-March, pp. 80–87.
3   Hollnagel, E. (2004). *Barriers and accident prevention.* Aldershot, UK: Ashgate Publishing Co.
4   North, D.M. (2000). Let judicial system run its course in crash cases. *Aviation Week and Space Technology*, May 15, p. 66.
4   Langewiesche, W. (1998). *Inside the sky.* New York: Random House, p. 228.

19 Making Your Safety Department Work

Being a safety department can be tough. Often you get pushed into roles like:[1]

- Being a mere arms-length tabulator of largely irrelevant or unusable data. For example, you may find yourself gathering numbers and producing bar charts about events or incidents every month that nobody really seems to use;
- Compiling a trail of paperwork whose only function is to show compliance with regulations or safety targets set by the company;
- Being a cheerleader for past safety records and remouthing how everything was better before (e.g. before deregulation, or before the new management);
- As a cost center whose only visible role is to occasionally slow down production;
- Getting excluded from organizational activities and decisions that affect the trade-off across production and safety goals;
- Delivering systemic safety recommendations while line management actually seems to feel that its responsibility after a mishap is to nail the people involved.

Granted, some companies *like* their safety departments to be in these roles. It means that the safety people are not going to ask difficult questions. They will not interfere with management work. They can conveniently be kept out of line decisions about which they would otherwise undoubtedly have bothersome opinions.

I remember working with one organization whose safety department was regarded solely as an intelligence gathering operation. Their role was to provide management with safety information (targets for how much safety information were even set up), and that was it: a mere bottom-up supplier. At the same time, management typically came down hard on operators who had been involved in incidents, often without waiting for more extensive investigation in what had gone wrong or why.

While the safety department looked on from the sideline, management went about squelching the very sources from which it owed its safety intelligence. With punitive measures meted out after each incident, operators got less and less inclined to report problems or events. This deprived the safety department of the data they needed to fulfill their targets.

So while management held up its one hand to receive safety data, it slapped operators around with the other. Nobody was apt to report about any safety-related issues any longer. The rather powerless, passive role given to the safety department was part of the problem, and the solution was to make it a much more active contributor to top-down management work—helping craft responses to incidents, debriefing people involved in them, and participating in other management decisions that affected trade-offs between safety and efficiency.

It is especially when production pressures are up, economic margins are small, and losses just around the corner, that safety departments may need to speak up. It is also precisely during these times that people in the organization could be less inclined to listen.

Escaping the Passive Safety-Supplier Trap

So how can you escape the role of passive number-supplier, cheerleader of the past, nagger, or mere cost center? Here is what you should expect:

- **Significant and independent resources** (both human and monetary). These resources must be independent of production or financial fluctuations for two reasons. First, safety monitoring and vigilance may actually have to go *up* when economic cycles go *down*. So you may need the resources most when the organization can least afford them. Second, a safety department may contain part-time operators. These have to be shielded from production ebbs and flows, because safety concerns often accelerate when production pressures go up (production pressures that could simultaneously rob the safety department of its people).
- **A constructive involvement in management activities** and decisions that affect trade-offs across safety and efficiency, as well as involvement in managerial actions in the wake of incidents. If you remain on the sideline as a supposed "impartial" department, you will see the basis for your work be whittled away. You must be ready to do battle for your data and the safety culture they feed and represent.

- An **end to weekly or monthly or quarterly targets**. Safety is not something that the safety department produces, so targets make little sense. They also favor quantitative representations (bar charts) of supposed safety issues, rather than qualitative intelligence on what is actually going on—*when* it is going on (not because the week or month is up).
- A continued **grounding in operational reality**. Having only full-time safety people can make your department less effective, as they can lose their idea (or never had one) of what it is to operate at the sharp end, and what the shifting sources of risk out there may be. Make sure you do not let one professional group (e.g. doctors and pilots, versus nurses and flight attendants) dominate your department, as this will skew your safety intelligence and ways of dealing with it.
- That said, just being a practitioner does not qualify people to be members of a safety department. You must expect **education** of staff members in safety management, incident/accident investigation, report writing, presentation, and so forth. Without this, you can quickly get surrounded by happy amateurs whose well-meaning efforts contribute little to the quality and credibility of the safety function.
- An organizational location of your department in a **staff position**, not somewhere down the line. You should be as independent of (yet not insensitive to) current economic and political concerns as possible. Also, you should have direct access to relevant organizational decision-making levels without having to pass through various levels of non-practice oriented middle management.

And here is what you have to show in return:

- **A sensitivitity to legitimate organizational concerns** about production pressures and economic constraints. Showing that you care *only* about safety can quickly make you irrelevant, as no organization exists just to be safe. They are there to attain other goals, and you are an inextricable part of that. Without achieving production goals for example, everybody may as well pack up and go home. Even the safety people, as soon there will be no more operation to monitor.
- **Real safety intelligence**, as opposed to distant tabulations of statistical data. This means that you have to show how safety margins may have been eroding over time, and where the gap lies between what causes problems in the operation and what management believes causes problems. This will require you to go beyond statistics and come up with persuasive intelligence.

For example, you may have to lay bare the innards of an incident that can reveal to management how it could have been misinterpreting the sources of operational risk and their own role in shaping that risk.

A confidential reporting system is a key source of safety-related information for most organizations. While practitioners themselves may not see some of their trade-offs or decisions as an "incident" (which an outsider perhaps would), and will thus not report about those, a confidential reporting scheme can still help you get details about events which would otherwise elude you completely.

Mike O'Leary and Nick Pidgeon[2] showed how much the formal and confidential reports about the same event can diverge, and how little actionable, interesting information the formal report often contains. Those formal reports often constitute (or underlie) the sort of target-based management safety information churned out by safety departments. They do not create much insight at all.

Here is an example of the difference: Formal report: "On final approach at 2000 feet AGL the first stage of flaps was selected. Flaps failed with no flap movement. A decision was made to go around and hold while preparing the aircraft for a flapless approach. Flapless approach completed on runway XX".

Part of the confidential report: "During the go-around, I was distracted by concern for proximity of high ground as the clearance was non-standard. 'Gear-up' was called. I was about to select it up when Air Traffic Control called again. After the call, I continued the after-take-off checks as if the gear was up. Neither of us realized it was still down for some five minutes."

The confidential report touches on issues that never even made it into the formal one. These are issues that reveal additional exposure to risk in a situation like the one described, and how crews try to manage that risk. The effects of this mix of unexpected system failure, workload, and ATC distractions are never mentioned in the formal report, and thus would never make it into wider organizational consciousness. With the confidential report in hand, organizations can think about investing in broader meaningful countermeasures (other than just fixing the flap problem on that one aircraft), such as influencing the design of go-around procedures (at this particular airport and others), workload management, double-checking and crew coordination.

A Concerned Outsider Who Knows the Inside

You should strive for a role as a concerned outsider who understands the inside. For your deparment to see how safety margins may be eroding over time, you have to monitor the organization's understanding of itself. Without knowing it, the organization may have chosen to operate closer to the boundaries of safe performance than it would actually like to.

The organization will not see this itself. It typically interprets the inevitable signs of operating close to the boundaries in localized, unthreatening ways (e.g. inattention, complacency of operators, or any other "human error"). In other words, you have to show the organization how it may be misinterpreting itself. It may be misleading itself into believing that it has a narrow human error problem, which is actually the effect of problems much deeper inside the operation, created by the organization itself.

You want to help the organization revise or reframe its assessment of risk. The challenge may be to convert the managerial perspective: from seeing human error as a source of risk, to human error as an effect of trouble deeper inside the system. This includes helping the organization rethink the effectiveness of its countermeasures against what it perceives as its biggest risk. You may find fertile ground for your efforts: organizations can occasionally feel hopeless and helpless when human errors just don't go away, however many reprimands, rules or regulations they issue.

Informed, Independent, Informative and Involved

A good safety department is about managing a conflicted role. It is about balancing multiple pressures and goals that can quickly work in opposite directions. David Woods, who followed the travails of the NASA safety organization through two Space Shuttle accidents, offers how a safety department has to be simultaneously informed, independent, informative and involved. These can be difficult to reconcile, but it is in a precarious balance among these properties that a safety department can do its best work:

- **Informed** about how the organization is actually operating (as opposed to just taking distant numerical measurements). This requires the safety department to have its feet firmly planted in the operation, and its ear to the ground (e.g. via confidential reporting).
- An **independent** voice that can challenge accepted interpretations of risk (e.g. a "safety attitude problem" perceived by management may actually be the expression of an entirely different problem deeper inside the organization). Having a strong voice at the top is not enough: the safety department must be tightly connected to actual work processes and their changing, evolving sources of risk.

- **Informative** in terms of helping reframe managerial interpretations of safety threats and helping management direct investments in safety when and where necessary. It is important that the safety department has access to relevant managerial decision-making channels, but also that it finds new ways to transmit its insights. In-depth analysis of critical incidents and meaningful discussion of this during meetings where all relevant stakeholders are present with ample time, could be effective in recalibrating how an organization looks at itself.
- **Involved** constructively in organizational actions that affect safety. Activities such as responding to incidents, or granting waivers, should not be done without consulting the safety department, as all of these activities can substantially influence downstream operational safety.

The leadership of one organization got rather concerned when its safety department announced it needed to be "independent". Not because this would allow them to discover and say things that could make the leadership uncomfortable (which has happened too, in other organizations), but because it was thought to make the safety department less involved in what was going on in the organization. Much discussion, and the inclusion of expert practitioners in the safety department, helped clarify the nature and importance of its "independence".

Staying independent but involved is difficult, as involvement may compromise independence and vice versa. Staying informed (what is going on?) but informative (how can we look with a fresh perspective on what is going on?) also conflicts. Learning about safety requires close contact with the risk inherent in operational practice, as well as distance for reflection.

Resolving these conflicts is impossible, but managing them consciously and effectively is feasible. It requires that the safety department is large enough to carry various perspectives, that it knows it has to be seen as a constructive participant in the organization's activities and decisions that affect the balance across production and safety goals, and that it stays firmly anchored in the details of daily practice to know what is going on there.

Feedback to Line Management

Crucial to making a safety department work is its link to line management. In the wake of an incident, line management often sees multiple responsibilities, which may conflict. It needs to help the organization learn something from the mishap, and improve where necessary. But this can quickly take the form of reprimanding or retraining individual operators, as this satisfies other goals as well (e.g. showing that you are doing something about the problem).

New rules for European air traffic safety management stipulate how line managers, in the wake of an incident, need to check the competency of the controller involved in the incident. While having competent personnel on the job is indeed a line responsibility, doing such a check as a response to an incident sends the wrong signal and sets the wrong tone. It risks institutionalizing the Bad Apple Theory. The system is fine, and the incident would not have happened, if only the competency of this or that air traffic controller would have been adequate. Of course, you can wonder if line managers will admit at all that their personnel were not properly qualified or competent for the job anyway—as this reflects badly on them too.

One way to turn this around is for line managers to consider the psychological consequences of an operator having been involved in an incident. What happens to the person's self-confidence, for example? If concern around "competence" centers around such post-incident questions, rather than insinuating that the person in question may not have been competent for the job all along, there is a way forward and a possibility for improving safety.

Finding personal shortcomings

The search for personal shortcomings, however, is not foreign—neither in management debriefings of personnel involved in incidents, nor even in formal investigations. Rules for investigations in some domains specify that you have to look into the 24-hour or 72-hour history of the persons involved in the mishap (supposedly to see whether something in the human components was already bent or broken). In the best cases, this is merely a process of "elimination", a background check to rule out longer-standing or sudden vulnerabilities that were particular to the people in question (e.g., fatigue, medical issues). But such hunts can easily fuel the Bad Apple Theory: suggesting that the failure is due to personal shortcomings, either temporary or long-running, on the part of the people involved in it.

Remember the submarine that collided with the fishing vessel, sinking it? The admiral testifying about his subordinate commander's performance explained how on a ride a year before he noticed how the commander perhaps did not delegate enough tasks to his crew because of his own great talent. The admiral had had to remind the commander: "let your people catch up".[3]

Hindsight seriously biases your search for evidence about people's personal shortcomings. You now know where people failed, so you know what to look for, and with enough digging you can probably find it too—real or imagined. We are all very good at making sense of a troubling event by letting it be preceded by a troubling history. As Chapter 3 has shown, we construct plausible, linear stories of how failure came about once we know the outcome, which includes making the participants look bad enough to fit the bad outcome they were involved in. Such reactions to failure make after-the-fact data mining of personal shortcomings not just counterproductive (sponsoring the Bad Apple Theory) but probably untrustworthy.

This means that immediate responses to incidents should probably not be left to those with line responsibilities for the people involved. Debriefings of personnel can too quickly take the form of a performance review, in which shortcomings can easily be highlighted in the glaring beam of hindsight. Defensive mechanisms will be triggered and people involved may get more concerned with mitigating the personal consequences of the incident. Instead, post-incident debriefings should be an opportunity for everybody to learn and improve. That means that information about them should be evoked in as open and non-judgmental an atmosphere as possible. The safety department, rather than line management, can play a role here.

Finding systemic shortcomings

Putative shortcomings of individual operators can instead be used as a starting point for probing deeper into the systemic conditions of which their problems are a symptom. Here are some examples:

- From their 72-hour history preceding a mishap, individual operators can be found to have been fatigued. This may not just be a personal problem, but a feature of their operation and scheduling—thus affecting other operators as well.
- Training records may sometimes reveal below average progress or performance by the people who are later caught up in a mishap. But it is only hindsight that connects the two; that enables you to look back from a specific incident and cherry pick associated shortcomings from a historical record. Finding real or imagined evidence is almost pre-ordained because you come looking for it from a backward direction. But this does not prove any specific causal link with actions or assessments in the sequence of events. Training records are a much more interesting source when screened for the

things that all operators got trained on, and how and when, as this explains local performance much better. For example, how were they trained to recognize a particular warning that played a role in the mishap sequence? When were they last trained on this? Answers to these questions may reveal more fundamental mismatches between the kind of training people get and the kind of work they have to do.
- Operators may be found to have been overly concerned with, for example, customer satisfaction. In hindsight this tendency can be associated with a mishap sequence: individuals should have zigged (gone around, done it again, diverted, etc.) instead of zagged (pressed on). Colleagues can be interviewed to confirm how customer-oriented these operators were. But rather than branding an individual with a particular bias, such findings point to the entire organization that, in subtle or less subtle ways, has probably been sponsoring the trade-offs that favor other system goals over safety—keeping the practice alive over time.

A commuter train derailed near Osaka, Japan in 2005, crashing into an apartment building. The accident killed at least 56 passengers and injured 440. The train had been running 90 seconds late, and was suspected of doing twice the speed limit for that section of the line, when five of its seven cars jumped the track.

Investigators focused on the inexperienced 23-year-old driver, who had been reprimanded for overshooting a platform once before during his 11 months on the job, and may have been speeding to make up for the delay. The Japanese Prime Minister called on officials to "respond firmly in order to prevent future accidents".

Trains in Japan, however, are known for running with such precision that riders traveling on unfamiliar lines can plot complex itineraries, secure in the knowledge that they will not miss connections because of delays. Accuracy and speed are key ingredients in running a successful commuter system in dense urban areas. The driver may have felt he was expressing the preferences and expectations of his entire society.[4]

Notes

1 The initial material in this chapter is based on Woods, D.D. (2006). How to design a safety organization: Test case for resilience engineering. In E. Hollnagel, D.D. Woods, and N. Leveson (eds), *Resilience engineering: Concepts and precepts*, pp. 296–306. Aldershot, UK: Ashgate Publishing Co.
2 O'Leary, M. and Pidgeon, N. (1995). Too bad we have to have confidential reporting programmes. *Flight Deck*, Vol. 16, pp. 11–16.
3 *International Herald Tribune*, March 14, 2001
4 *International Herald Tribune*, April 26, 2005

20 How to Adopt the New View

Getting *your* organization to adopt the New View may be challenging. At least, that is what you may feel. Suppose you sense that your organization's current approach to failure is not productive. You see how your organization easily blames "human error", and how it thinks that this can lead to meaningful countermeasures. And then you see it gets nowhere. You struggle to get your organization to change its perspective on how failure and success come about. Remember three things, then:

- **Embracing the New View is not an on-off switch**. It takes a lot of work, and is typically a difficult organizational, political process. You should expect a degree of conflict, acrymony, even resistance.
- **Growth towards the New View is not even**. Not all parts of the organization learn at the same pace. Some parts may lead the way while others lag behind. Efforts to embrace the New View will expose existing fault lines that separate vested interests and viewpoints in your organization.
- **Growth towards the New View is not regular**. Growth can accelerate and slow to a crawl. Regression into the Old View is even possible. Yet you may be able to identify stages in growth, which help you determine where you can expect your organization (or part of it) to go next.

Also, the ability of your organization to embrace the New View may be influenced by the general climate and attitudes towards failure in your industry or country. Where litigious approaches to human error prevail, or laws bar confidential reporting, it may be difficult to get people to embrace New View practices. Protective posturing will take precedence, as people and organizations will invest first and foremost in protecting themselves and their own interests. This obviously does not help learning or progress on safety—neither for your organization nor for anybody else in the industry or country. These cases call for higher-level discussions (industry-level, political) about how to respond to failure more meaningfully.

The Stages of New View Acceptance

As in all growth, it is possible to distinguish certain stages in an organization's gradual embrace of the New View. You can see these stages like stages in any spiritual growth. They lead to greater consciousness, to an opening of minds.

The stages are "ideal types". They may correspond roughly to what your organization (and its key people) are going through. But given that growth toward the New View is typically irregular, partially event-driven, and unevenly distributed across an organization, many routes of New View maturation are possible. The stages may help you identify some stations along that multifarious route, so you know what could come next.

The stages of growth mark how an organization is making progress in learning about learning from failure. The more an organization embraces the New View, the more it will have learned about how it really can, and should, be learning from its successes and failures.

Stage 1: Crisis—paralysis of the Old View

When incidents keep occurring, and traditional measures (finding "human errors"; dealing with them with reprimands, tighter procedures, more technology) do not seem to work, an organization can become open to new perspectives. The first stage is often marked by desperation. People swat every fly ("human error") in sight, but the flies keep coming in because their kitchen is still messy and sticky. The paradox can be maddening: people think they are doing so much to deal with all those human errors, and yet they keep occuring. Perhaps even more than ever.

A major accident may similarly open a window of opportunity. It can create such a tension between how people thought the organization worked versus how it turns out to actually work, that they are forced to abandon old beliefs about what makes things safe or risky. The accident makes them see that risk is not the result of "human errors", and that maintaining safety is not a matter of containing such errors through ever more procedures, oversight or technology.

Crises in the management of safety can also be forced through factors as diverse as economic strain and cutbacks, internal organizational strife, an increase in efficiency demands or production pressures, or the adoption of new technology that brings unexpected side effects. These factors can all open up rifts in how different people think they should manage the organization, what they should emphasize or prioritize or expect (e.g. efficiency gains without any consequences for safety).

Stage 2: Disassembling Old View interpretations

The ideal result of a crisis from stage 1 is the realization that a human error problem is an *organizational* problem. This begins the process of distributing the responsibility for doing something about the problem.

When people begin to realize that human error is the effect of problems deeper inside the system, then this may guide them to the sources of those problems (e.g. goal conflicts) and how these systematically and predictably produce outcomes that, in an earlier era, got attributed to "human error". What they typically find is that operational work is surrounded and defined by possible hazards, complexities, trade-offs, and dilemmas. These problems get imported from many different corners of organizational life.

The challenge, of course, is to resist simply finding "human errors" deeper inside the organization, as this takes people back to before Stage 1. Finding out where responsibility for doing something about the problem can meaningfully lie, is forward-looking and oriented towards progress. It is not about placing blame.

In fact, stakeholders at this stage will begin to see that "human error" is not an explanation of failure, but simply an attribution, the result of social process (of which they were part). They begin to see that labeling human error as cause of trouble says nothing about the trouble itself, but about the people who do the labeling—about their beliefs in what exposes the system to risk. Embracing the New View is in large part about the ability to change these beliefs, so seeing how they are expressed in the first place is critical.

Stage 3: Freezing Old View countermeasures

An organization that has realized the uselessness of pursuing "human errors", does not necessarily know what to do next. In fact, the first stages of growth often create a sense of profound unease and managerial discomfort. People can no longer turn to traditionally satisfying ways of dealing with failure, as they now understand how counterproductive or paradoxical these are.

But other people, who have interpreted the crises differently, or have other stakes, may not agree. They can continue to exert pressure on managers or others "to do something" about the error problem.

A fruitful strategy, adopted by some organizations who are going through the growing pains of New View acceptance, is to put a freeze on all Old View countermeasures. Without precisely knowing what to do next, they will actually stop reprimanding people, not immediately turn to more procedures to solve the latest discovered gap in system operations, and resist accepting "just a little more technology" that would supposedly deal with the human error problem forever.

Such a freeze often generates surprising results, as those who favor traditional knee-jerk responses to the latest errors realize that witholding such reactions has no negative effect on the number or severity of incidents. In fact, there may be the reverse. Old View pundits will find out, empirically, that their calls for immediate action may have been counterproductive after all.

Stage 4: Understanding that people create safety through practice

A conversion to the New View turns on the insight that risk is not caused by people in otherwise safe systems. Systems are not basically safe, they are made safe through people's practice. The third stage can show how a freeze on traditional countermeasures (which presume that it is all about erratic people in otherwise safe systems) does not lead to increased risk. This helps open an organization up for the key insight: safety is not inherent in systems, only to be threated by operators-on-the-loose. People actually create safety through practice.

Getting to this insight is the major accomplishment that marks the fourth stage. Organizations will begin to see that people's practice has evolved to cope with the many hazards, complexities, trade-offs and dilemmas of their work. They have found ways to reconcile multiple conflicting goals to create safety in actual settings. Old View stories of failure will almost always end with some individual, in their head, their motivation, their (lack of) awareness or understanding. New View stories end in the world, in the system in which people worked, systems which people made work in the first place.

You will know that your organization has entered the fourth stage when it stops talking in terms of the widely known and rehearsed first stories (about how people committed "errors"). Instead, it will discuss multiple contributing factors, competing demands on operators and people throughout the organizational hierarchy, goal conflicts, the complexity of the processes people manage, and how they have adapted to usually succeed at this. These are the ingredients of the second stories about failure, they embody the notion of safety creation as an active product of efforts at every level of the organization.[1]

Stage 5: New View investments and countermeasures

Once your organization starts putting its money where its mouth is, you know it has entered stage 5. Here it backs up its new understanding—that safety is created by people in a thicket of multiple goals, divergent demands and complexity, and that failures are an inevitable by-product of the pursuit of success under those circumstances—by action, by investments.

New View countermeasures are investments in the system. People stop fighting symptoms. New View countermeasures, for example, will acknowledge goal conflicts and may help people manage them better. Not with a new procedure, but by helping people consider the post-conditions of points along the trade-off between safety and efficiency concerns, and giving them the courage to decide against common interpretations of what is risky or acceptable. The organization may also invest in a stronger safety department so that it can help itself get better calibrated about how closely it is operating to the margins.

A company that runs expensive trucks asked me how it could be that a couple of loaders, at the end of their workday, would knowingly use the wrong lift to offload a truck. Instead of using a conveyor belt that fits the side door of the truck nicely, they used a much broader elevator. As it came up, it softly pressed into the truck, causing damage that was not discovered until the next loading crew came on shift the day after.

"They should have known that this was the wrong lift!" exclaimed one of the managers. I replied that I was rather uninterested in finding out what they should have known, as this evidently did not play much of a role that evening. But I would like to find out why it made sense for them to do what they did.

It turned out that the end of a workday is not marked by the clock, but by the last truck coming in. As soon as that truck is offloaded, the crew can go home. Knowing that this was the last truck, and having heard that it would have no cargo in the front part, the crew had already packed away the conveyor belt, as well as other things they would no longer need that day. The store for these things was quite far from the parking platform. To their surprise, the crew did find a load in the front part of the last truck. They removed it using the lift that they still had around, even though the lift isn't made to fit the door to the front part.

The incentive structure for this work contributed in a critical way. If you say to people that they can go home after the last truck, then you can expect a flurry of assiduous housekeeping (i.e. packing everything but the bare bones away) before that last truck comes in. If they then are confronted by surprises—a load in a place where it is not supposed to be—improvisation with the tools at hand can also be expected. Systemic solutions included changing the incentive structure (let the clock determine when the workday is done and find economic ways of dealing with overtime) and providing better advance information about where loads are placed. Abandon the fallacy of a quick fix, I suggested to the manager. Look hard at the incentive structure and how you are contributing to people behaving in this way.

Stage 6: Learning how you are learning from failure

While your organization will never entirely complete its journey toward the New View, you can become more confident that it can stay on the road towards progress if the organization itself becomes interested in how it is learning from failure. In fact, an organization enters the stage in growth that "completes" the embrace of the New View when it is actively involved in finding out, and managing, how it learns from failure.

In other words, it is no longer just interested in revising the first-order activities (learning from failure by dispensing with Old View responses and replacing them with New View ones). In addition, it shifts its focus to second-order activities. It continually monitors *how* it is learning from failure. It asks what it can learn about itself from the way it is currently learning from failure.

Seeing how an organization grows in New View maturity, and helping it move through subsequent stages, should not just be the activity of select individuals, a few side-line guardians of growth (e.g. a safety staff). The extent to which an entire organization (i.e. *all* relevant stakeholders, up and down the line) actively considers how it is learning from failure—large and small, on a daily basis—says something powerful about the organization's safety culture. Organizations with a strong safety culture continually try to understand how they learn about safety and respond to threats and opportunities at all levels of practice. Having a safety culture involves being able to calibrate which stage of growth the organization (or parts thereof) is in or going through. It means meta-monitoring, or meta-managing: monitoring how you monitor safety; managing how you manage safety.

This is the essence of maturation: realizing that there is no final destination, no terminal. Rather, your organization becomes capable of monitoring how it grows, where it is in its growth. It becomes able to identify, articulate and influence how its activities and responses are contributing to that growth.

Language helps you mark the growth

As organizations grow through these stages, you will notice a gradual transformation in language too. A New View embrace is necessarily accompanied by the adoption of a new vocabulary about safety and risk. You will see a transformation:

- **From** a language of individual people, who "could have" or "should have" done something different from what they did; from a language of illusory

psychological afflictions (complacency, inattention, loss of situation awareness, loss of effective crew resource management),
- **To** a language of situations, systems, and structures; to a language that captures the context in which people work, including the constraints on, and opportunities, for individual action.

The distribution of organizational members' language across these two possibilities tells you something about where they are in their maturation toward the New View. It helps you determine which stage of growth they may be in.

Don't Get Sidetracked by False Calls

In the wake of failure, or other pressure for change, people can issue calls to action that seem very much like the New View. But in reality these are misguided. Following them, and believing that they will lead to progress on safety, is like building a house on sand, or sowing seeds on the rocks. These calls are, for example:[2]

- Demands for increasing the general awareness of the issue among the public, media, regulators and practitioners. For example, people may claim that "they need a conference …". Increasing general awareness may be a good first step, but is no guarantee for getting the New View any traction. Without actual work inside organizations, and the accumulation of empirical evidence that the New View helps people create safer operations, there is no sustainability and the awareness raising may have been a waste of time.
- Insistence that progress on safety can be made easily if some local limitation is overcome ("we can do a better job if only …"). Overcoming these local limitations may be clad in New View systemic rhetoric, but it may resemble sticking a finger in the hole of the dyke to keep the floodwaters out. It could work, but only for a little while.
- Calls for more extensive, more detailed, more frequent and more complete reporting of problems ("we need mandatory incident reporting systems with penalties for failure to report"). The gathering of data is not the analysis of data, and is certainly a long distance from doing anything useful with it. It can not be an end in itself and calls that focus only on such measures are deceptive. Also, "failures to report" cannot be remedied by an even more punitive stance. Willingness to report is a complex function of people's own perceptions of risk and organizational culture.

- Calls for more technology ("We need computer order entry, automated load sheets, bar coding") that creates an additional layer of defense. These calls too, can be cloaked in New View oratory, but are really an expression of the Old View (the system needs to be better protected against erratic, unreliable people).

These calls are all Old View countermeasures in some disguise. They tend to attribute the cause of problems to narrow, proximal factors ("human error" of some kind, by somebody). They may seem like attractive countermeasures to failure. But they actually lead to sterile responses that will limit learning and foreclose opportunities to really improve.

Don't Expect it to be Easy

If they really are growing, organizations will begin to see that an engagement of the New View not only makes sense, but also generates a return on investment in terms of safety: the New View starts promoting itself. Most people will not readily give up such newly explored and discovered consciousness by pretending that it doesn't exist, by suddenly closing their minds. Yet, in some cases, such retardation is possible. When put under enormous strain (after a big failure, with significant consequences, even potentially for one's carreer), people and organizations have been known to regress into name-calling ("human error!"), as they got pressured into "doing something quickly".

The resilience of your organization's New View embrace depends on how mature the organization and its people have become; where they started their growth, and how many of the stages of growth they have passed by what number of the organization's key members.

The debate between Old View and New View interpretations is not just about two perspectives, about an intellectually quaint exercise in comparing worldviews. It is often about vested interests and dire stakes at the heart of any organization. Battles to learn about safety typically exposes existing organizational stress, and can amplify it. As Richard Cook reminds us, virtually all components of the New View expose the underlying rifts, disagreements and mixed character of the organization itself. These conflicts, though unsettling, are crucial to learning about safety. They are the source of the opposing demands and resource limitations that determine and constrain practitioners' and managers' ability to create safety.[3]

Notes

1. Cook, R.I., Woods, D.D., and Miller, C.A. (1998). *A tale of two stories: Constrasting views of patient safety*. Chicago, IL: National Patient Safety Foundation.
2. Woods, D.D., and Cook, R.I. (2002). Nine steps to move forward from error. *Cognition, Technology and Work*, 4, 137–144.
3. Cook, R.I., Woods, D.D., and Miller, C.A. (1998). *A tale of two stories: Constrasting views of patient safety*. Chicago, IL: National Patient Safety Foundation.

21 Reminders for in the Rubble

This chapter wraps together some the most important lessons from the Field Guide. It gives you a summary of the New View and its implications. It presents you with five reminders each about:

- Your own organization and the nature of safety and risk in it;
- What to think about when investigating human error;
- Doing something about your human error problem;
- How to recognize Old View thinking;
- How to create progress on safety with the New View.

Whatever you try to understand about human error, do not forget to take the point of view of the person inside the situation (see Figure 21.1)

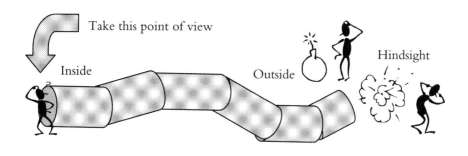

Figure 21.1 If you want to understand human error, see the unfolding world from the point of view of people inside the situation—not from the outside or from hindsight.

Your Organization and the Roots of Error

1 Your organization is not basically or inherently safe. People have to create safety by putting tools and technologies to use *while* negotiating multiple system goals at all levels of your organization.

2 The priorities and preferences that people express through their practice may be a logical reproduction of what the entire organization finds important.

3 Human error is the inevitable by-product of the pursuit of success in an imperfect, unstable, resource-constrained world. The occasional human contribution to failure occurs because complex systems need an overwhelming human contribution for their safety.

4 So human error is never at the root of your safety problems. Human error is the effect of trouble deeper inside your system.

5 It also means that human error is not random. Human error is systematically connected to features of people's tools, tasks and operating environment.

What to Think of when Investigating Human Error

1. Nobody comes to work to do a bad job. This is the local rationality principle. People do what makes sense to them at the time given their focus of attention, their knowledge and their goals (which may well be the organization's goals, stated or unstated).

2. Human error is not the cause of failure, but the effect. So human error, under whatever label (loss of situation awareness, complacency, inadequate crew resource management) can never be the conclusion of your investigation. It is the starting point.

3. Explaining one error (e.g. operator error) by pointing to another (inadequate supervision, deficient management, bad design) does not explain anything. It only judges other people for not doing what you, in hindsight, think they should have done.

4. To understand what went on in somebody's mind, you have to reconstruct the situation in which that mind found itself.

5. There is no such thing as *the* cause of a mishap. This is like looking for *the* cause of not having a mishap. What you deem causal depends on your accident model.

Doing Something About Your Human Error Problem

1. Do not get deluded by the fallacy of a quick fix. "Human error" problems are organizational problems, and so at least as complex as your organization.

2. Reprimanding supposed Bad Apples (errant operators) may seem like a quick, rewarding fix. But it is like peeing in your pants. You feel relieved and perhaps even nice and warm for a little while. But then it gets cold and uncomfortable. And you look like a fool.

3. If you are a manager or supervisor, you cannot expect your employees to be more committed to safety than you yourself are, or appear to be.

4. Problems result from your organization's real complexity—not some apparent simplicity (e.g. somebody's innattention).

5. Do not expect that you can hold people accountable for their errors if you did not give them enough authority to live up to the responsibility you expect of them.

Recognizing Old View Thinking

1 Old View thinking sees human error as the major threat to basically safe systems. Unreliable, erratic people undermine systems of multiple defenses, rules, procedures and other safeguards.

2 Old View thinking will try to count and categorize errors, and endeavour to get the human error count down from its stubborn 70%. It assumes that safety, once established, can be maintained by monitoring and keeping people's performance within prespecified boundaries.

3 Old View thinking will (unsuccessfully) try to revert to more automation, tighter procedures, closer supervision and reprimands to control erratic human performance.

4 During downsizing, budget trimming and increased production pressures, Old View thinking will misinterpret human errors as a source of trouble, when they are likely to be the inevitable downstream consequences of trying to do more with less.

5 Old View thinking judges rather than explains human performance. It uses language such as "they should have…", "if only they had…" and "they failed to…". But by saying what people should have done, you don't explain at all why they did what they did.

Creating Progress on Safety with the New View

1 To create safety, you don't need to rid your system of 70% human errors. Instead, you need to realize how people at all levels in the organization contribute to the creation of safety and risk through goal trade-offs that are legitimate and desirable in their setting.

2 Rather than trying to reduce "violations", New View strategies will find out more about the gap between work-as-imagined and work-as-done—why it exists, what keeps it in place and how it relates to priorities among organizational goals (both stated and unstated).

3 New View thinking wants to learn about authority-responsibility mismatches—places where you expect responsibility of your people, but where their situation is not giving them requisite authority to live up to that responsibility.

4 You know your organization is maturing towards the New View once it actively tries to learn how it is learning about safety. This means your organization is calibrating whether its strategies for managing safety and risk are still up-to-date.

5 Every organization has room to improve its safety. What separates a strong safety culture from a weak one is not how large this room is. What matters is the organization's willingness to explore this space, to find leverage points to learn and improve.

Index

accidents
  analytic trace 125–33
    bottom up 125, 127–8
    top down 125, 129–33
  causes 17
  CFIT 98
  models 81–92
  probable causes 78–80
accountability
  models 201–3
  without authority 200, 228
  see also responsibility
after-the-fact-worlds 29–30
Air Ontario incident 79–80, 178
analytic trace
  bottom up 125, 127–8
  conversation analysis 127–8
  investigations 124–31
  top down 125, 129–33
authority
  and accountability 200, 228
  constraints on 198–200
  responsibility, mismatches 196–203
automation
  and coordination 153–4
  traces 98
automation surprises
  human error 84, 114–15, 152–4
  new technology 149

Bad Apple Theory 1–14
  features 2
  futility of 9–10, 228
  investigations as 6
  popularity 10–12
  sidelining 216–18
behavior
  and human error 71–2
  and process 111–14
Bhopal incident 88
blame game 4, 188–9
blunt end, vs sharp end 59–60
Boeing 747 incident 5, 45–6, 51–2
buggy knowledge, and human error 145–6

causality
  construction of 73, 75–80
  pressure to discover 73
  'probable causes' 78–80
  simplification of 25
cause-consequence equivalence 24
cause-effect relationships 84
CFIT (Controlled Flight Into Terrain) accidents 98
Challenger incident 164, 169
cherry-picking, data 29, 33–5
Clapham Junction railway incident 23
cognitive
  consequences, computerization 150–51
  fixation 135–9
Columbia incident 83–4, 90–91
compartmentalization, human error 187–88

complacency
    and human error 119, 120
    meaning 131–3
complex systems 17
    and human error xi
computers
    cognitive consequences 150–51
    dumbness 150
    and human error 149–51
    interface problems 150
    keyhole problems 149–50
context
    data in 29–38, 123–4
    and hindsight 29
conversation analysis, analytic trace 127–8
coping strategies, stress 141–3
counterfactuals
    examples 48–51
    and hindsight 40–41
    human error 39–44, 195
    language use 39, 48–54
countermeasures, New View 218–19
CRM (Crew Resource Management) 119, 120, 122, 124, 125, 128, 129–30, 131

data
    availability 32
    cherry-picking 29, 33–5
    in context 29–38, 123–4
    events identification in 114–16
    and hindsight 31–2
    human factors 93–9
    leaps of faith 120–23, 124
    micro-matching 29–33
    observability 32
    out of context 29–37
    overload, new technology 149
    performance, recordings 97–8
DC-9 incident 126
DC-10 incident 82, 178

debriefings
    aim 94–5
    inconsistencies 96
    questions to ask 95–6
demand-resource mismatch 141–2
disaster relief work, variant images 167
dumbness, computers 150
dynamic fault management 139

epidemiology, model 81, 82–3, 87–90
error
    problem of defining 66–8
    see also human error
ETTOs (Efficiency-Thoroughness Trade-Offs) 198
events, linearity of 25
evidence, and hindsight 35–6
explanation, vs indignation 45–57

face saving 11
failure
    avoidance of word 42–3
    as by-product 17–18
    false calls 221–2
    and hindsight 23–4, 40–41
    learning from 220
    perspectives on 26
    reactions to 21–3, 60
        counterfactual 21
        distal 60
        judgmental 22
        proximal 22, 60, 63
        retrospective 21, 22–3
    system overview 59–61
    see also human error
fatigue
    causes 144
    and human error 143–5
    and performance 144
fault management, dynamic 139
Fitts, Paul 15
folk models, human factors 121–2, 131–3

goal conflicts
  causes 169–70
  coping strategies 168–71
  identification 170
  NASA 166
  and production pressure 164–71

hard fixes 189–92
  see also quick fixes
hindsight
  and context 29
  and counterfactuals 40–41
  and data 31–2
  and evidence 35–6
  and failure 23–4, 40–41
  investigation, effect on 23
  as Old View 28
  and simplification 24–5
  ubiquity of 27–8
  and viewpoint 24–5, 29
human error
  active 88
  automation surprises 84, 114–15, 152–4
  and behavior 71–2
  blame game 4, 188–9
  and buggy knowledge 145–6
  categorization 68–9
  causes of 3, 17, 73–80, 135–58
  cognitive fixation 135–9
  compartmentalization 187–8
  and complacency 119, 120
  and complex systems xi
  and computers 149–51
  as coping strategy 67–8
  counterfactuals 39–44, 195
  definition, difficulties of 66–8
  and fatigue 143–5
  hard fixes 189–90
  and inert knowledge 145–7
  labels 119–23
  language use 48–57
  latent 88
  location 68–70

mechanical failure, distinction 73, 74–5
  and mind-matter divide 72
  nature vs nurture 201
  need to understand 14
  and new technology 9, 19, 147–52
  and non-compliance 119, 120
  normalizing 186
  Old View, vs New View xi
  opportunities 18–19
  as organizational problem 159, 195, 226
  plan continuation 140
  and procedural adaptations 154–8
  quantification 65–72
  quick fixes 183–94, 228
  reconstruction 70–72, 227
  reporting, avoidance of 186–7
  stress 140–43
  as symptom x, 4, 18, 226
  viewpoint 225
  see also Bad Apple theory; failure; New View; Old View
human factors
  data 93–9
  essence 13
  folk models 121–2, 131–3

indignation
  as Old View 45
  vs explanation 45–57
interface problems, computers 150
International Civil Aviation Organization 176
investigations
  analytic trace 124–31
  as Bad Apple Theory 6
  and hindsight, effect of 23
  key points 227
  and New View 18–19
  and Old View 5–6
  and personal shortcomings 211–12
  perspectives 26
  purpose 5

keyhole problems, computers 149–50
knowledge, inert, and human error 145–7
   see also buggy knowledge

Ladbroke Grove train incident 59, 61
language use
   counterfactuals 39, 48–54
   human error 48–57
   New View of human error 220–21
   Old View vs New View 48–57
leaps of faith, data 120–23, 124
learning
   from failure 220
   nature of 184
   openness to 183
local rationality principle 13, 48, 61–3, 185

mechanical failure, human error, distinction 73, 74–5
micro-matching, data 29–33
mind-matter divide, and human error 72
mode errors, new technology 148, 150
models
   accident 81–92
   accountability 201–3
   epidemiological 81, 82–3, 87–90
   human factors 131–3
   predictive use 81–2
   procedures 156–7
   purpose 81
   sequence-of-events 81, 82, 83–7
   see also folk models; systemic accident model
Murphy's Law 164, 165

NASA
   goal conflicts 166
   safety department 209
nature vs nurture, human error 201
new technology
   automation surprises 149
   complexities 148
   data overload 149
   display confusion 148, 149
   and human error 9, 19, 147–52
   mode errors 148, 150
   non-coordination 148–9
   operational pressures 151–2
   workload 149
   see also computers
New View 1, 3–4, 15–20, 222
   adoption 215–22
      stages 216–20
   countermeasures 218–19
   and investigations 18–19
   language use 220–21
   procedures model 157
   research insights 17–18
   responsibility-authority mismatches 197
   and safety progress 19, 230
   technology role 147–8
non-compliance, and human error 119, 120

Old View 1–2, 216–18, 229
   hindsight as 28
   indignation as 45
   and investigations 5–6
   procedures model 157
   responsibility-authority mismatches 197
   safety progress 7
   shortcomings 12–14
   thinking 229
omnipotence, illusion of 11–12
operational
   pressures, new technology 151–2
   vulnerability 184–5
organizations, and human error 159, 195, 226
Osaka train incident 213

PBL (Problem-Based Learning) 146
performance, and fatigue 144

performance data, recordings 97–8
plan continuation 140
practice, and procedures, mismatches 8, 30–31, 159–64
predictions, recommendations as 173–4
'probable causes' 78–80
procedures
  adaptations, and human error 154–8
  application 155–6
  enforcement 7–8
  models, opposing 156–7
  and practice, mismatches 8, 30–31, 159–64
process, and behavior 111–14
production pressure, and goal conflicts 164–71

quantification, human error 65–72
quick fixes 183–94
  examples 185
  human error 183–94, 228
  see also hard fixes

recommendations
  countermeasures identification 179–80
  high-end 175
    participant involvement 179
    problems with 177–9
  low-end 174–5
  as predictions 173–4
  problems with 180–81
  SMART implementation 174
  trade offs 175–6
reconstruction, human error 70–72, 227
regression, stress 142
responsibility
  authority, mismatches 196–203
  safety 196–203
  see also accountability

safety
  auditing 192–4
  confidence in 192–4
  creation 16, 65, 218, 226
  culture, elements 172
    inconstancy of 163–4
  levels 190–92
  monitoring 192–4
  negative attitudes to 2
  personal responsibility 196–203
  policies, written 191
  progress
    New View 19, 230
    Old View 7
  supervision 191
  trade offs 16–17
  whistleblowers 193
safety departments 205–13
  constraints on 205–6
  education role 207
  line management, feedback to 210–13
  NASA 209
  outputs expected 207–8
  proactivity 206–7
  requirements 206–7, 209–10
  role 208–9
  safety intelligence provision 207–8
  targets, inappropriateness of 207
sensemaking, dynamics 136–8
sequence-of-events
  countermeasures 84–7
  model 81, 82, 83–7
sharp end, vs blunt end 59–60
shortcomings
  personal, and investigations 211–12
  systemic 212–13
simplification
  of causality 25
  and hindsight 24–5
situation awareness, loss of 135–6
SMART implementation, of recommendations 174
Space Shuttle
  Challenger incident 164, 169
  Columbia incident 83–4, 90–91
standards, imposition of 32–3

stress 140–43
  coping strategies 141–3
  regression 142
  and time perception 143
  triggers 141–2
  tunneling 142–3
Swissair 111 incident 54–7
Swissair MD-11 incident 78–9, 86
symptom, human error as x, 4, 18, 226
system overview, failure 59–61
systemic accident model 81, 82, 83, 90–92
  advantages 92
  basis 91
systemic shortcomings 212–13

technology *see* new technology
time perception, and stress 143

timeline
  communication
    high-resolution 101, 108–11
    low-resolution 101, 104–5
    medium-resolution 101, 105–8
  events in 114–16
  limitations 102–3
  types 101
tunneling, and stress 142–3

ValuJet flight 592 incident 198–9
viewpoint
  and hindsight 24–5, 29
  human error 225

whistleblowers, safety 193
work, variant images of 167–8, 193
workload, new technology 149